D1423442

FRANCIS BACON
Discovery and the Art of Discourse

FRANCIS BACON
Discovery and the Art of Discourse

LISA JARDINE
Research Fellow, The Warburg Institute,
University of London

CAMBRIDGE UNIVERSITY PRESS

Published by the Syndics of the Cambridge University Press
Bentley House, 200 Euston Road, London NW1 2DB
American Branch: 32 East 57th Street, New York, N.Y.10022

Library of Congress Catalogue Card Number:

ISBN: 0 521 20494 1

First published 1974

Printed in Great Britain
at the University Printing House, Cambridge
(Brooke Crutchley, University Printer)

CONTENTS

PREFACE

The present work was developed from my Ph.D. dissertation for the University of Cambridge, during the tenure of a Senior Research Fellowship at the Warburg Institute, London. I extend grateful thanks to the Director, staff and fellows of the Warburg Institute for many stimulating and informative discussions which have greatly contributed to my understanding of Francis Bacon, and to my general appreciation of the classical tradition.

I acknowledge with gratitude the patient and critical advice of my Ph.D. supervisor, Dr R. R. Bolgar. I am indebted to Dr N. Jardine for advice, particularly on the tradition of scientific demonstration, and for help with German texts. I am deeply grateful to Dr M. B. Hesse, Dr D. P. Walker and Dr C. B. Schmitt for their careful and critical reading of the manuscript, and for general advice and encouragement. I have had useful discussions with Dr D. S. Brewer, Dr G. K. Hunter, and Dr B. W. Vickers, and have received advice on points of detail from Professor P. Rossi, Dr G. E. R. Lloyd, Dr D. S. T. Clark, Dr W. H. Donahue and Miss J. Weinberg.

My thanks are due to the librarian and staff of the University Library, Cambridge, the British Museum Library, the Warburg Institute Library, London, the Francis Bacon Library, Claremont, California, University College Library, London, and the library at Gorhambury House, St Albans, for making their facilities generously available to me.

Throughout the present work I have cited from J. Spedding, R. L. Ellis, and D. D. Heath, *The Works of Francis Bacon* in fourteen volumes (London, 1864–74). As far as I know, no subsequent work on Bacon's manuscripts has produced any improvements or significant additions to Spedding's text. My own examinations of the contents of the library at Gorhambury House; the collection of commonplace books in the Ogden collection of Bacon texts (some of which may have been made for Bacon by various of his secretaries, though I am inclined to doubt this), now in the library of University College London; the British Museum additional manuscripts; and the contents of the Francis Bacon Library, Claremont, California, yielded nothing which Spedding had not already noted. Occasionally translations in this edition are unclear, or show lack of awareness of technical terms current in the period, but on the whole I have tried to clarify these passages in my own glosses, and have not tampered with the text.

References to volume and page of this edition are given in square brackets in the text for convenience. In general I have cited the English translation of a passage, where a translation is given. In order to help the reader to identify the work in which a passage occurs from the volume and page citation I give here the location in the Spedding edition (London printing) of the main works which I have used: *Sylva Sylvarum* [II, 333–680]; *New Atlantis* [III, 127–66]; *Valerius Terminus* [III, 217–52]; *Advancement of Learning* [III, 261–491]; *Plan of the Great Instauration* [IV, 5–33]; *Novum Organum* [IV, 39–248]; *Parasceve* [IV, 251–71], *De Augmentis* [IV, 275–V, 123]; *History of Life and Death* [V, 215–335]; *History of Henry VII* [VI, 24–245]; *Essays* (1625) [VI, 373–518]; *Essays* (1597) [VI, 523–34]; *De Sapientia Veterum* [VI, 689–764]; *Colours of Good and Evil* [VII, 75–92]; *Apophthegms New and Old* [VII, 123–86]; *Maxims of the Law* [VII, 313–87].

1974 L.A.J.

Introduction

Bacon called his *Novum Organum* a *logic*. In the *De Augmentis*, the work in which he maps out his provisional scheme for a revised classification of knowledge, he assigns to the *Novum Organum* or Interpretation of Nature the pivotal position which was traditionally given in such classifications to Aristotle's Organon. It is to provide the universal tool for a complete understanding of all those things which are accessible to human reason; all intellectual activities which are not subject to its analysis belong either to the realm of divine inspiration or to the subordinate disciplines which merely manipulate existing knowledge in a plausible manner. Moreover, both in the *Novum Organum* and in the *De Augmentis* Bacon presents his Organon specifically as a remedy for methodological shortcomings of conventional Aristotelian logic teaching; that is, he presents it as providing solutions for problems raised and (to his mind) unsatisfactorily dealt with by contemporary dialecticians.

In the present work I take seriously Bacon's own view of the branch of knowledge to which his most striking intellectual innovation belongs, and follow up the consequences of his placing a revised logical tool at the focus of his programme for the advancement of learning. To do this I look at his works as a whole against the background of the sixteenth-century dialectic handbooks from whose simplified treatment of Aristotle's Organon anyone but a professional philosopher (which Bacon was not) received his logical training. The dialectic handbook not only gave him a grounding in simple syllogistic and associated argumentative techniques, but also placed the study of dialectic within learning as a whole, and defined the areas of its application.

I hope to show that this background provides a helpful viewpoint from which to consider Bacon's writings over the wide range of disciplines to which he contributed. We have works by Bacon on scientific method, practical science, law, pedagogic theory, English history, myth interpretation, as well as occasional literary works. Each of the branches of learning to which these works individually belong was assigned by Bacon a precise place within his system for the growth of knowledge. And this means, as I shall show, that each work stands in a precise relationship to the general study of intellectual processes, the study comprising both the new Organon and the conventional dialectic (which Bacon demotes to become a subsidiary on a par with rhetoric) on which all knowledge depends. It is thus possible to anticipate what techniques of argument and strategy of composition Bacon will, if he is to be consistent, employ for a work within a given field.

In humanist-inspired education programmes like the one which Bacon himself followed at Cambridge, conventional dialectic played just this central, organising rôle in the teaching of particular arts and sciences. If its prominence is particularly noticeable in Bacon's works it is because he held particularly strongly the view that all search for knowledge is dominated and controlled by certain universally applicable organising procedures. As I shall show, *method*, the study of such procedures had, during the sixteenth century, become an increasingly central topic in dialectic. As a topic in dialectic, *method* in theory covers both the investigatory procedures which reveal new knowledge – what I have chosen to call *discovery* – and procedures for selecting and arranging existing information for purposes of communication and instruction – what in the period came to be called the *art of discourse*. For Bacon a distinction between these two activities is crucial for the progress of knowledge. His theoretical work on scientific method takes the discovery/presentation dichotomy as its point of departure. His creative and didactic works exploit the artificiality of literary conventions in direct contrast to the natural constraints imposed on procedures of investigation.

I should make it clear at the outset that my aim is to provide a consistent rather than a revolutionary reading of Bacon's works. By reconstructing the intellectual backcloth against which Bacon posed the particular methodological questions which engaged his attention I believe we are led to appreciate the originality and ingenuity of his own solutions, and are restrained from reading back into his work a concern with issues which have subsequently been judged crucial for the development of scientific and philosophical method. To take a single example, the fact that Bacon calls his scientific method an *induction* and rejects the pure deduction of syllogistic as inappropriate to scientific enquiry has prompted historians of science to hail him as anticipating modern views on the rôle of problematic induction in science, and to pillage the *Novum Organum* for supporting evidence. In fact, in the context of sixteenth-century treatments of method, an *induction* is generally accepted to be the method employed in the formation of the universal concepts and necessary truths on which scientific knowledge (for them) depends, and Bacon's own induction is closely related to suggested Aristotelian solutions to the problem of the accessibility of first principles. This does not, in the event, detract from the originality or interest of Baconian induction, but neither does it suggest any *prima facie* reason for attributing to Bacon a strict anticipation of a modern view of scientific method. Indeed, if we respect the context in which Bacon himself places his discussion, the *Novum Organum* is a richer source of understanding of the development of modern science than if we wilfully impose on him our own preoccupations and expectations.

I shall illustrate the usefulness of this approach by sketching revised readings of the *Novum Organum* itself, the *De Sapientia Veterum* and the *Essays*. I hope that this study may provide an interesting case history for other historians of ideas, and show how an uneasy alliance between Aristotelian dialectic and experimental science produced a logic of scientific discovery which on a cursory inspection appears to have all the features necessary to make it a forerunner of modern scientific method. It should provide a warning against judging Bacon according to our modern canons, and at the same time, I hope, stimulate further interest in the peculiar rôle played by renaissance Aristotelianism in the emergence of modern thought. And as far as students of literature are concerned, I hope that the intimate relation between Bacon's 'scientific' and his literary ideas may stimulate them to explore further the interactions in the period between what have become for us two separate cultures.

What then was the scope of dialectic in the period in which Bacon wrote? In the mediaeval curriculum dialectic was studied alongside grammar (the study of the structure of language) and rhetoric (the study of the embellishment of utterance) as part of the *trivium*, the tripartite study of the arts or practical activities concerned with language. Grammar provides the bricks, well-formed linguistic units, out of which all utterance is constructed. Dialectic builds these linguistic units into complete edifices of reasoned speech. Rhetoric graces the exterior of the completed construction. The rules of the trivium subjects are thus fundamental to all human intellectual activity, since it is assumed that this is conducted verbally.

In addition to material straightforwardly concerned with correct Latin usage the mediaeval grammar handbook contained important philosophical discussions of the correspondence between the structure of language and the structure of the mental and physical worlds. Although these discussions are often highly simplified they contribute to a view of language as providing a perfect map for process and change in nature. Dialectic is then seen as analysing natural relations as embodied in discourse, and manipulating language to gain insight into the natural world.

Within such a framework dialectic is considerably broader in scope than the subject which we now call 'logic' (the formal study of inference). At least in theory it covers all study of rationally constructed discourse. It therefore takes over from Aristotle's Organon not only those parts which we regard as dealing with points in formal logic, but also much which we would assign to linguistics or philosophy of language. The traditional treatment of dialectic, based on Porphyry and Boethius, included discussion of the relations between subject and predicate in a sentence, the kinds of predicate, the kinds of sentence, and techniques for forming and criticising definitions. The dialecticians of the later middle ages added to this *logica antiqua* a *logica moderna* in which sophisticated discussions of meaning, reference, tense and modality are used to tackle problems (notably in theology and metaphysics) which cannot be dealt with using only the logical apparatus of the *logica antiqua*.

These later logicians take it for granted that linguistic techniques may legitimately be used to extend the power of dialectic, and they take it for granted that the dialectician's model of language ought to capture the subtleties of ordinary usage. It is noticeable that in this later period there is considerable overlap between the scholarly discussions of grammarians and dialecticians on points of linguistic detail.

Sometime early in the fifteenth century there was a gradual reorganisation and revision of the dialectic handbook. Whereas in the earlier period the teaching manual for dialectic condensed and simplified the sophisticated arguments of professional logicians, a new type of handbook emerges which was explicitly angled towards the teaching of a general course to a non-specialist. Those responsible for the change were, as one might expect, the advocates of a broad, classically based education, the educational reformers commonly referred to as 'humanists'.

In the restyled handbook the broad scope of dialectic, implied in the traditional tag 'dialectic is the art of arts and science of sciences, giving access to the principles of all disciplines' ('dialectica est ars artium scientia scientiarum ad omnium methodorum principia viam habens') is preserved. But the philosophical justifications for this breadth of scope are discarded, and with them the detailed mediaeval treatment of terms and the relations between them, and the linguistic subtleties of the *logica moderna*. Instead there is increasing emphasis on the *practical* virtues of dialectic as a means of organising and displaying subject matter in any discipline. These dialecticians do not present the rules of dialectic as the universal rules of discursive reasoning, but consider it primarily as a tool for the teaching of curriculum subjects.

It is in the context of display and presentation that the theme of *ordo* or *methodus* appears in these revised dialectic handbooks, the treatment of the rules for the extended organisation of subject matter not as a chain of formal inferences, but as a table of related questions or topics. The term '*methodus*' is deliberately chosen to suggest an analogy between chains of inference in more formal logic and in mathematics (to which the term was already applied) and these less rigorous, but more useful, procedures for ordering.

Like the revisionists of the dialectic handbook Bacon required of his logic that it be useful in practice. But his criterion of usefulness was efficacy in the extension of knowledge, not efficacy in its presentation. He insists that no amount of displaying and arranging of received pedagogic material will yield the fundamental principles of knowledge. His 'logic', the Novum Organum, is his replacement for the reformed dialectic and its methods, and at various points in it he tackles specific questions which he believed the dialecticians had overlooked, or had failed to handle adequately. The crucial issue for Bacon was whether a logic could be devised which would provide a universally appropriate model for the procedure of scientific discovery – the deriving of principles too fundamental to be arrived at by syllogistic proof techniques from existing data. Throughout his work Bacon stresses the, to his mind basic, distinction between 'discovery', that is, the investigation of the unknown by way of his new logic, and 'invention', that is, the selection of received assumptions about the natural world as premises for argument or for display. In a sense this means that Bacon's challenge to the dialecticians rests on a misunderstanding. It was never the intention of the humanist reformers to offer a logic of discovery; for them dialectic was the key to teaching, and clear and eloquent expression in all fields of knowledge.

Bacon's work has never been discussed against a dialectical background. To do so is to concede that he was intellectually a provincial. On the continent the same period produced competent discussions of the extended *methodus* of formal reasoning in science, in the context of commentary on Aristotle's *Posterior Analytics* and *Physics*. These attempt in various ways to formalise intellectual processes of acquisition of knowledge as chains of inferences, starting from descriptions of immediate sense perceptions. I maintain that Bacon was ignorant of such discussions although they have often been assumed to be direct antecedents of his inductive method. Those who have looked at Bacon's inductive method first and foremost as a *scientific* method, on the other hand, have turned to the writings of earlier natural philosophers, and looked there for methodological precedents. Once again, I suggest that this is to look too far afield, and (in the detailed comparisons with the scientific procedures of Grosseteste and Roger Bacon) to assume far too great an expertise and scholarship from Bacon. I suggest that some of the problems which have dogged Bacon scholars dissolve if one regards Bacon as a well-educated English gentleman with a good (but not scholarly) grounding in the curriculum subjects, and with a remarkably clear grasp of precisely the limitations of that education as the basis for any growth of understanding of the natural world.

I begin with an account of the way in which dialectical methods, and controversy about the scope and validity of their applications, emerged from the fifteenth- and sixteenth-century reforms in dialectic teaching. In the remainder of the book there is a primary division into discussion of Bacon's treatment of acquisition of knowledge and discussion of his treatment of presentation of knowledge. Each of these sections itself falls naturally into two parts, one concerned with Bacon's *aspirations* in the field, the other with his attempts to put his methods into practice. Inevitably there is a discrepancy between these two aspects of Bacon's writings. His theoretical writings are, as I shall show, closely linked with contemporary discussions within dialectic; my task has been to draw out the relevant themes from Bacon's works, and to indicate the associated topics in dialectic. In putting these theories into practice, however, Bacon tended to continue, sometimes perhaps quite unconsciously, existing traditions. Here I have tried both to show how he developed existing genres, and to point out the ways in which his theoretical ideas were modified in practice. Both types of analysis are heavily *textual*. On the whole I am more concerned to give a coherent reading of Bacon's *oeuvre* than to take issue with particular critical interpretations in the recent secondary literature.

It is, of course, in a sense question-begging to describe a selected background before describing those works whose details are supposed to be explained by precisely that background; but to appreciate correspondences between Bacon's discussions and those of contemporary dialecticians, one does need a reasonable grasp of the key issues in dialectic. In the later chapters which deal in some detail with particular themes in Bacon's writings, my claims that Bacon's obsession with procedure of discovery and methods of presentation is in fact a key to the understanding of his works, and that this obsession can be related to earlier controversy about the nature and correct application of dialectical methods, may be put to the test.

One may, however, begin by indicating some of the more obvious reasons for looking at Bacon's works against this particular background, starting with direct evidence based on terminology in Bacon's works and historical and biographical facts. Bacon regards as fundamental to his scheme of knowledge the division of human reasoning into two 'moments', 'invention' and 'judgment'. In all intellectual activity we first collect together our perceptions or experiences (invent), and then assess them, and decide to act upon them (judge). Division of dialectic as a whole into invention and judgment, a tradition deriving from Cicero, is the hallmark of the reformist dialecticians. And within this dichotomy of intellectual activities, Bacon's characterisations (in the *De Augmentis* in particular) of features of existing logic match very closely the treatments of the same topics in popular dialectic handbooks of the period. Amongst these we might list the 'invention' which explicitly takes first principles on trust from the science or art under discussion; the use of 'topics' or 'places' as the basis for selection of material; the 'judgment' by induction (which Bacon calls 'puerile') which considers a few striking affirmative cases and ignores counter-examples or the incompleteness of the enumeration of affirmatives; the characterisation of treatment of the syllogism, including 'ostensive' syllogistic proof and proof '*per incommodum*'; the brief discussion of the four traditional types of logical demonstration; the use of the sophistical fallacies as a caution against unsound reasoning. Even the three rules for checking the scientific validity of premises or principles, now generally associated with Ramus' dialectical innovations, are to be found discussed under 'demonstration' in standard contemporary handbooks in the very form in which Bacon uses them. In fact, as I shall show, as far as Ramism is concerned, whilst clearly aware of Ramist controversy (he alludes to it as having 'moved a controversy in our time' [III, 403], and criticises some specifically Ramist tenets), Bacon's only direct borrowing appears to be his use of the term 'axiom' atypically as Ramus uses it, for any proposition used as a premise for argument.

The reformed dialectic manual based on Rudolph Agricola's

pioneer handbook, the *De Inventione Dialectica libri tres* (to which I shall have repeated occasion to return), was specified for the Cambridge syllabus at the time at which Bacon attended the University. The 1560 statutes for Trinity College (which Bacon attended) likewise name Agricola, and in one version Seton's elementary manual based on Agricola and Melanchthon, as basic texts for dialectic teaching.

We know from the personal account books of Archbishop Whitgift, who was in charge of Bacon's studies at Trinity, that he customarily bought the dialectic manuals of Seton and Caesarius for his students. One may surmise that for Whitgift's students, at least, these two were standard dialectic texts. This view is also supported by surviving inventories of the stocks of contemporary Cambridge booksellers, in which Seton is particularly prominent, and in the students' personal booklists preserved amongst the probate records in the Cambridge University Archives. These two latter sources show a remarkable preponderance of dialectic manuals (five or six occurring in a single student booklist; several copies of upwards of ten titles on the shelves of booksellers), all conforming to the type of 'reformed' manual I have described.

Booklists and inventories once again confirm that Ramist texts were widely circulated at the time at which Bacon attended Cambridge (1573–5). We know that Bacon's tutor was hostile to Ramism, and that he did not purchase Ramist texts for his own students. The dispute between two Cambridge dons (Everard Digby and William Temple) about the comparative merits of Ramist and Aristotelian methods, shortly after Bacon left Cambridge, suggests that there was considerable interest in dialectical methods in the university at the time.

There is in addition an argument from negative evidence. Rawley tells us that Bacon was 'no plodder upon books' [I, 12]. His account of Bacon's reading habits conveys an impression of a voracious, but unsystematic reader, and a glance at the footnotes to the Spedding edition of the scientific works shows that in natural philosophy, despite his conventional parade of classical sources and recondite information, Bacon appears to have made use of a fairly limited range of popular encyclopaedic works. For example, though it would be nice to believe, as some critics have suggested, that the experimental bias of Bacon's scientific method is in part derived from the works of Roger Bacon, the bulk of his references to Roger Bacon appear to be based only on a short and thoroughly unexperimental excerpt in a popular collation of magical texts, and many of the key scientific questions to which he gives attention ('whether a vacuum exists?', 'whether the moon is the cause of the tides?', 'whether principles are pairs of contraries?') correspond to standard topics or *quaestiones* set to students for debating as part of their initial training. It would be equally plausible to suggest that all Bacon's allusions to atomism and atomistic doctrines can be traced to a few popular sources. This is very much the situation when we consider the often noted 'Aristotelian' presuppositions in Bacon's writings on procedure and method. Tempting as it is to see Bacon's works (as some critics have done) as a development of the sophisticated reassessment of Aristotle's Organon by the schools of Oxford, Paris and Padua, and hence to invoke a vast scholastic background, it is in fact possible to account both for Bacon's 'Aristotelian' assumptions and for his 'anti-Aristotelian' polemic in terms of the content of the dialectic manual, and contemporary polemical discussions of dialectical method.

Such evidence does not justify the claim that the dialectic handbook provides more than one of a host of backgrounds against which one might consider Bacon's works. To establish my stronger thesis it needs to be shown that in tackling some of the questions which are central to his writings as a whole, Bacon takes up issues which belong specifically to dialectic, and in particular to dialectical theories of method, and debates these in their own terms.

In Bacon's writings on procedure of discovery we shall see that such a correspondence is not difficult to establish. The *Novum Organum* is explicitly presented as a reaction to the approach of the 'dialectici', and to a logic which takes the first principles of its field of application on trust. And we have Rawley's often-quoted remark that 'whilst [Bacon] was commorant in the university, about sixteen years of age' (that is, at a time when the dialectic handbooks of authors like Seton and Caesarius provided his detailed knowledge of logic and Aristotle's Organon) 'he first fell into the dislike of the philosophy of Aristotle' [I, 4]. (Since his tutor Whitgift was an old-school

Aristotelian it is unlikely that he prompted or encouraged Bacon's mistrust of the old Organon.)

The dialecticians, as Bacon frequently complains, regarded the discussion of derivation of principles as lying outside the scope of their work. On the whole they were content to invoke the Aristotelian tag that the principles of any discussion are drawn from the knowledge of experts in the particular field in which discussion takes place, whilst the principles common to all fields are immediately seen to be true by the discerning mind. When pressed, however, they conceded that the ultimate basis for belief in principles is an induction. The source for this is again Aristotle, in the *Posterior Analytics.* Bacon points out that the induction which the dialecticians actually discuss in their handbooks is an imperfect form of inference which is clearly unsuitable for the task of deriving sound principles, but he accepts the fact that some form of induction is in order. His own method is an attempt to perform such an induction, that is, to infer the general rule from the particulars in which it inheres, by another route. What is striking, and supports the claim that his knowledge of the logic of discovery was restricted to what he learnt from the dialectic handbook, is that he appears to be unaware of the sophisticated discussion by continental commentators on the *Posterior Analytics* of just what Aristotle intended by the induction which yields indemonstrable principles.

The universal rules of *philosophia prima,* which even for Bacon remain beyond any induction, are precisely the first principles recognised as common to all sciences by the dialecticians (Melanchthon discusses them particularly clearly in his *Erotemata Dialectices*). Even Bacon, that is, believes that some principles are of such generality that they are simply recognised as true by an astute observer of the natural world (principles like, 'when equals are added to equals the results are equal', or 'the whole is greater than any of its parts'). And *philosophia prima* is not the only component of his system taken over uncritically from the dialectic handbook. Like the dialecticians he takes it that the basic movements of all intellectual activity are those 'up' from what is prior or better known to us to what is prior or better known in nature, and 'down' from what is prior or better known in nature to what is prior or better known to us. He also apparently retains a rather naive view of explanation of particular phenomena once principles have been established. Scientific principles are essential definitions or the immediate consequences of essential definitions, and reasoning based upon them is apparently (to judge from a letter written to Baranzano) conventionally syllogistic.

In the case of Bacon's theory and practice of presentation of existing knowledge (as opposed to the derivation of unknown principles) the points to be made about their connection with the mid-century dialectic manual are rather more simple and general. In the first place, it is possible to point out some general implications of the reforms in dialectic teaching for the theory of literary composition in the period.

The sixteenth-century student learnt both the theory and practice of literary composition within the framework of his basic training in the trivium. At a practical level this course was designed to teach him the rudiments of composition by direct imitation – to this extent one might designate the Cambridge arts course as 'humanist'. Both in his latter years at school and during the preliminary years of his university education, heavy emphasis was laid on close, critical reading of approved works by classical authors, identifying and classifying the techniques of composition and formulae of embellishment used, and reproducing the best features of these in prose exercises modelled on what had been read. To aid this learning procedure reading was done in conjunction with model-books of various types. Some of these listed individual ornamental devices, some collected together model passages, from a single genre (model letters, for instance), from an author or group of authors (*flores*), or from specified types of discourse (*progymnasmata*). The student's own personal guide to composition was his commonplace book; in this he recorded useful phrases, effective arguments and particularly successful rhetorical devices noted in the course of his reading, for his own future use (these commonplace books also served incidentally to provide teacher or tutor with a check on his pupil's reading progress).

As far as *theory* was concerned, the principles of composition were by this time taught first and foremost from the 'new look' dialectic handbook. This fact has, I think, been insufficiently emphasised in the past. It was, as I have said, a key tenet of the reformed dialectic that the rules of dialectic are basic to all understanding of the organisation of discourse. Valla and Agricola's rejection of the more abstruse and technical aspects of mediaeval logic as irrelevant to the educational programme which they favoured was prompted by a belief that the rules of dialectic should provide all and only such technical knowledge as was necessary to the educated man for the composition of lucid and elegant prose. And by the time that Melanchthon and Caesarius wrote their rhetoric manuals, these were openly presented as supplementary and subsidiary to the dialectic manual, and as presupposing detailed knowledge of dialectic. In the introduction to his *Rhetorica*, Caesarius stresses the fact that grammar and dialectic are 'prior' to rhetoric in the analysis and composition of discourse. Grammar gives the rules for well-formed utterance, dialectic the rules of combination and organisation of terms and propositions, and these are the basis for competent use of language. Rhetoric adds the often *ad hoc* rules for discourse which *moves* an audience, and in effect Caesarius' manual brings together all such hints on composition, embellishment and delivery, derived from Cicero, Quintilian and the pseudo-Ciceronian *Ad Herennium*, as do not fit the format of the dialectic handbook. However subtle the techniques for moving an audience's emotions by style, choice of attitude towards a subject, or embellishment with tropes and schemes, these authors assume that the foundation for all successful utterance is *rational*, and that dialectic provides the body of rules for rational discourse.

At Cambridge at least, in the mid-sixteenth century, the detailed theoretical instruction in literary composition which formed part of the basic arts course was provided by the dialectic handbook. This was supplemented by extensive personal reading of recommended authors in conjunction with rhetoric handbooks and model-books which took the content of the dialectic handbook for granted. And whether or not a student's particular training incorporated a discussion of 'method' (which would depend to some extent on the interests of the tutor who directed his study), this instruction focused on the ability to put together extended arguments on a specified theme, and to lay out a body of material on a given topic in a coherent and orderly way.

Precisely because the reformed dialectic handbook was designed to eliminate needless and misleading distinctions between a contrived, formal treatment of language peculiar to logicians and natural oratory, Valla and Agricola drew heavily on the works of Cicero and Quintilian for the sections of their treatment of the 'art of discourse' specifically concerned with the laying out of complete discourses, or entire academic disciplines for teaching purposes. The rather general and literary remarks of these Roman authors are in striking contrast to the more technical pronouncements of Aristotle on ordering of subject matter. This tends to give fifteenth century discussion of *ordo* or *methodus* a distinctly 'rhetorical' flavour; and indeed, the passages from the *Ad Herennium*, for instance, had previously been used as the basis for a cursory treatment of *dispositio* of complete discourses within rhetoric (for example by Trapezuntius). However, the dialectical discussions of principles of organisation should not be regarded as mere reallocation of an old topic in rhetoric. As we shall see, the dialecticians made conscious attempts to adapt the discussion to pressing problems of presentation, in particular those associated with teaching, which had not been the concern of classical or mediaeval rhetoricians.

Bacon's views on presentation very clearly belong to the development which I have outlined above. He subordinates rhetoric to dialectic, and characterises rhetoric as something overlaid upon discourse composed according to dialectical rules. As I shall also show, Bacon's theoretical treatment of the presentation of existing knowledge falls squarely within the tradition of discussion of method in the dialectic handbook, and he recognises the connection between this discussion and earlier discussions of rhetorical *dispositio*.

Whatever shortcomings Bacon found in the dialectic handbook as a source for the rules of operation of language as a scientific instrument, he continued to uphold it to be 'very properly applied to civil business and to those arts which rest in discourse and opinion' [IV, 17]. As a guide to literary composition, that is, rather than as a blueprint for scientific discovery, he was satisfied with the dialecticians' treatment of the 'art of discourse'. To avoid misunderstandings which from his point of view had misled some dialecticians into claiming their presentational methods as tools of discovery, Bacon proposed to set up a separate study of the 'method of discourse', specifically concerned with 'the rules of judgment upon that which is to be *delivered*' [III, 403; my emphasis]. What is remarkable about Bacon's brief theoretical discussion of 'methods' in the *De Augmentis* is the wide range of presentational conventions he is prepared to regard as tools for controlling and directing the reader's attention and understanding. He brings together under this heading such disparate techniques as the contrived question and answer format of the scholastic teaching manual, the diagrammatic schematisation of the Ramist manual and the epitome, the use of classical myths as an allegorical framework for ethical and philosophical precepts, and the general use of comparison and similitude to match abstruse ideas to the aptitudes of an unscholarly audience. He regards method as strategy, to be selected for effective teaching and persuading of an audience, and hence dismisses the entire dialectical controversy about the absolute merits of particular dialectical methods, in which it was assumed that some specified method might uniquely match the process by which a student acquired his knowledge from a teacher. All method is for Bacon to a greater or lesser extent an artifice for convincing presentation.

Given the extent to which teaching of practical composition dominated a student's early training, with the minimum of regard for theory and the maximum emphasis on imitation of classical authors, it is more difficult to pin down the ways in which Bacon's own works reflect his attitude towards dialectic and the theory of composition which dialecticians taught. One would expect Bacon's prose to reflect his training in the identification and use of the conventional figures of rhetoric, and several authors have in the past investigated this aspect of Bacon's writing. However, I maintain that there are important features of Bacon's prose composition which resist description in terms of rhetorical formulae. In this context I shall make two general points. I shall show how his theory of method as strategy led him to choose particular literary forms for particular didactic purposes, to render inaccessible ideas perspicuous for a popular audience. And I shall suggest that the effectiveness of particular works derives from a determination to exploit the available instruments for precise and convincing argument with complete freedom to make essentially literary points. Stripped of any mystique as a tool of discovery, or as determining a unique teaching method, dialectic is made available for opportunistic use. Discourse may be moulded according to the rules of dialectic with just the imaginative gusto with which it can be decorated with tropes and schemes. And this exploitative use of dialectic produces a tautness in discourse which commands sustained attention. Ben Jonson paid tribute to these powers of presentation in Bacon's public speeches: 'His hearers could not cough, or look aside from him, without loss. He commanded where he spoke; and had his judges angry and pleased at his devotion. No man had their affections more in his power' [I, 13–14].

1

Dialectic and method in the sixteenth century

The development of dialectic in the sixteenth century is essentially a development within a textbook tradition. It is characteristic of textbook writers that they tend to revise existing texts; textbooks are rarely completely innovatory. Probably this is because in all periods the community of schoolteachers has been a rather conservative and overworked body, unprepared for startling deviations in the standard presentation of their subject. It is true of the dialectic textbook tradition as of others that if one compares superficially the contents and treatment of topics in mediaeval and renaissance texts there is no striking discontinuity.[1] It is however possible, in the case of dialectic, to take a broader view, and to relate apparently minor changes in content to a gradual shift in attitude towards the subject of dialectic as a whole, and the position its study occupies within the teaching programme of the seven liberal arts. It is this which is important in the present context, since what I am concerned with is precisely what Bacon saw as the rôle of dialectic within a system of knowledge, and why he found fault both with its stated aims and its achievements. For this purpose what is important is how the subject is defined, how it is introduced to the student, how related to other subjects within the curriculum, and how the conventional content is grouped and emphasised. This is what I shall try to sketch here.

Dialectic is one of the three arts of the *trivium*, the other two being grammar and rhetoric. Together with the four *quadrivium*

[1] Compare for example late-fifteenth-century editions of Peter of Spain's *Summulae Logicales* (composed about 1246) and, say, editions of Caesarius' *Dialectica* (first published in 1532) commented in the 1580s, or Seton's *Dialectica* (first published in 1545) in Carter's commented edition of 1572. W. Risse has given a comprehensive survey of such internal developments, borrowings and omissions for the sixteenth century. See W. Risse, *Die Logik der Neuzeit*, Band I (Stuttgart, 1964).

subjects (arithmetic, geometry, astronomy and music) and the three philosophies (natural philosophy, ethics and metaphysics) these make up a classification of knowledge which was established as fundamental in the thirteenth century.[1] This classification provided the basis for an arts education in the tradition of the Latin West of late antiquity; in particular (since we are here concerned first and foremost with Bacon) it provided the foundation for the four year arts course for the degree of Bachelor of Arts at the universities of Oxford and Cambridge.[2] Roughly, the trivium embraced those studies whose object was language and its use, the quadrivium embraced mathematics and the main fields in which mathematics was applied, while the three philosophies employed the technical training of the trivium and quadrivium in the study of the accidents of natural phenomena, human behaviour, and *first philosophy* (the study of the general concepts and principles underlying all scientific or philosophical investigation), respectively.

The trivium occupied a central and key position within this scheme of learning. It taught the student how to be articulate, to handle language and argument with considerable subtlety, and to express his views in a convincing fashion. All these skills were prerequisite for the rest of the curriculum, based as it was to a large extent on examining and testing readings of set texts, debating controversial issues, defending a particular school of thought in some specialist field, and so on. And at a deeper level, the trivium was seen as crucial because it systematised language itself, and the way thoughts are expressed in language. All other arts and sciences are in a very basic sense secondary to such a system.

[1] The standard scheme of the three parts of philosophy which the Middle Ages inherited from antiquity consisted of logic, physics and ethics. Logic was also one of the seven liberal arts, so that the two schemes overlapped. This did not bother the earlier Middle Ages, since they had no conception of philosophy as a separate teaching subject. Metaphysics acquired its place with natural and moral philosophy (and with logic) in the thirteenth century when philosophy began to be taught at the universities.

[2] The names most closely associated with twelfth- and thirteenth-century developments of this classification as the basis for the arts curriculum are those of John of Salisbury, Dominic Gundissalinus, Robert Kilwardby, Albertus Magnus and Thomas Aquinas. On mediaeval classifications of knowledge and their developments see L. Baur, 'Dominicus Gundissalinus, *De Divisione Philosophiae*: herausgegeben und philosophiegeschichtlich untersucht nebst einer Geschichte der philosophischen Einleitung bis zum Ende der Scholastik', *Beiträge zur Geschichte der Philosophie des Mittelalters*, ed. Baeumker and von Hertling, Band IV, Heft 2–3 (Münster, 1903); J. A. Weisheipl, 'Classification of the sciences in medieval thought', *Mediaeval Studies*, 27 (1965), pp. 54–90.

Within the trivium, the relative emphasis placed on the three subjects, grammar, dialectic and rhetoric, varied according to the preoccupations of those who developed and taught them. At the end of the fifteenth century, at the point at which the present survey starts, Latin grammar was taught very much as a preliminary to dialectic (using, for instance, Priscian's *Institutiones* in conjunction with a 'modern' philosophical grammar like Alexander of Villa Dei's *Doctrinale* of *c.* 1199). The parts of speech, tenses and persons of the verb, and so on, were justified in the course of their exposition as corresponding when correctly formed to the real relations between states of things in the external world. In the course of his training in grammar the student was introduced to much of the technical terminology essential to dialectic.[1] Rhetoric was taught as supplementary to dialectic, and as a more circumstantial study of choice of material, formal composition of an oration, ornamentation and delivery. Dialectic, which provided the technical apparatus for the analysis of language, dominated the trivium programme.

Peter of Spain's *Summulae Logicales* (1246) was widely used as the dialectic textbook in universities until the 1520s, and remained in use in some places as late as the seventeenth century; it gives us a clear picture of the way in which the student was introduced to the subject which was the pivot of his university studies. The treatise is made up of seven tractates, the first six of which cover the content of the key Aristotelian texts and their standard commentaries which formed the basis for instruction in dialectic at university level. These were Porphyry's *Isagoge*, Aristotle's *De Interpretatione* and *Categories* (together called the *logica vetus*, and considered as dealing with terms and propositions); Aristotle's *Prior Analytics*, *Topics* and *De Sophisticis Elenchis*, and Gilbert de la Porrée's *Liber de Sex Principiis* (together called the *logica nova*, and considered as dealing with argumentation). The seventh tractate, the *parva logicalia*, is a mediaeval addition to Aristotelian dialectic (known as the *logica moderna*). Although Peter of Spain placed this at the end of his treatise, the *parva logicalia* was understandably considered by later commen-

[1] For a particularly clear and illuminating account of the relation between grammar and dialectic in the fifteenth century, and later developments in the two fields, see T. Heath, 'Logical grammar, grammatical logic, and humanism in three German Universities', *Studies in the Renaissance*, 18 (1971), pp. 9–64.

tators like Eckius to belong to the treatment of *terms*, and hence strictly to follow tractate three. Peter of Spain gives no treatment of Aristotle's *Posterior Analytics*, the treatise on scientific or demonstrative reasoning.[1]

This course is openly presented as a technical training in the analysis and classification of terms and propositions, and in forms of inference, as tools for disputation and public debating. The techniques are above all *critical*. The successive classifications and schematisations provide the student with a comprehensive programme which enables him to check and countercheck an opponent's arguments, starting with the well-formedness of his premises, and proceeding through the consistency of his reasoning to the acceptability of his conclusions. For this purpose the successive classifications are considered as convenient check-lists which ensure that nothing has been overlooked, rather than (as in Aristotle, apart from the *Topics*) primarily for their equivalence to natural classifications of the phenomena with which language deals.[2] Dialectic is presented as the study of the rules for sifting and classifying information expressed formally in language, and Peter of Spain particularly emphasises the interdependence of grammatical and dialectical analyses of language, as related studies of well-formed utterance.

Tractate I of the *Summulae* (which corresponds to Aristotle's *De Interpretatione*) begins with a brief semantic grounding for dialectic, which serves to underline the position adopted in the text as a whole, namely that dialectic is a study of *language* and its usage, rather than of concepts and thought processes. This is followed by a description of the various types of proposition, with clear instructions on how to identify them by their form. The basic proposition is the *categorical* proposition, a simple affirmative statement like 'man is an animal'. Tractate II (based on Porphyry's *Isagoge* with Boethius' commentary) handles the *predicables*, the supposedly natural classes into which universal terms fall by virtue of their relation to the subject of a proposition. In relation to 'red', 'colour' is the *genus*; in relation to 'colour', 'red' is a *species*; in relation to 'red campion' (amongst

[1] On the *Summulae* see J. P. Mullally, *The Summulae Logicales of Peter of Spain* (Indiana, 1945) (however, the translation of tractate seven included in this work is unnecessarily opaque); also Heath, art. cit., pp. 41–5, for an intelligent sketch. There is now a good critical edition of the text by L. M. de Rijk, *Tractatus, called afterwards Summule logicales*... (Assen, 1972).

[2] Heath makes this point, art. cit., p. 43.

the flowers called 'campion'), 'red' is a *difference* (it differentiates it from other species in that genus); in relation to 'mail van', 'red' is a *property* (it invariably accompanies the occurrence of a mail van, but a mail van of another colour is conceivable); in relation to 'this handkerchief', 'red' is an *accident.* Peter of Spain presents the predicables, as Aristotle does in the *Topics,* as the most effective and scientific means of demolishing an opponent's argument, that is, as a debating tool.[1] An opponent's argument is undermined if we can detect a faulty formulation according to the predicables. And this will be done by referring to what we already know about the classes into which the objects of discourse fall. This means that dialectic as presented by Peter of Spain is concerned with the *content* of an argument, unlike modern formal logic which is concerned only with specifying the forms of valid argument irrespective of content.[2]

Treatise III (based on Aristotle's *Categories* and Gilbert de la Porrée's *Liber de Sex Principiis*) gives a treatment of the ten classes or types to which individual predicates may be allocated by virtue of the respect in which they describe. We may say of 'a buttercup' that its flower 'is two centimetres in diameter', that it 'is yellow', that it 'belongs to the genus *ranunculus*', that it 'attracts insects', that 'I pluck it', 'in summer', 'in the hedgerow', that it 'grows upright', that it 'is in full bloom'. Each of these possible descriptions of 'this buttercup', falls into one of the *categories* or *predicaments, substance* ('this buttercup') *quantity, quality, relation, action, passion, when, where, position, state.* And under particular circumstances one or other of these kinds of description may be called upon appropriately in support of an argument. To complete the treatment of terms there follows an account of the *post-predicaments,* a classification of terms which fall outside the scope of the predicaments, and which include *priority, identity and motion* (roughly, the fundamental ways in which entities may be *related* amongst themselves).

The treatment of terms is followed by the treatment of argumentation. By the beginning of the fifteenth century noticeably

[1] Unlike his mediaeval predecessors he ignores sophisticated problems associated with the predicables, like the problem of universals.

[2] Risse fails sufficiently to appreciate this difference between the scope of renaissance dialectic and the scope of logic as we now understand it. He is led to devote a disproportionate amount of space to that part of dialectic which bears the closest resemblance to modern logic, namely syllogistic. See Risse, op. cit.

more emphasis was being placed on this part of dialectic than on the detailed analysis of terms. Johannes Eckius, in his commented edition of the *Summulae* of 1516, stresses that *argumentatio* ranks above analysis of terms, and is the 'primary subject' of logic.[1] Tractate IV (based on Aristotle's *Prior Analytics*) presents the student with a schematised version of the syllogism, and introduces a mnemonic for the figures and moods which was still in general use up to the nineteenth century.[2] The treatment of the syllogism occupies the focal position in the *Summulae*. Peter of Spain makes only passing reference to the subsidiary weak forms of inference, enthymeme, example and induction, which do not for him form part of the essential equipment for formal debating.[3] This is only one indication of the fact that Peter of

[1] *Joannis Eckii Theologi in Summulas Petri Hispani* (Augustae Vindelicorum, 1516), fo. Aiiir; see also Heath, art. cit., pp. 49–50.

[2] A syllogism consists of two premises and a conclusion derived from them. The two premises share a common term (the 'middle term'), which is eliminated in the conclusion (in the syllogism: 'man is a rational biped, Socrates is a man, therefore Socrates is a rational biped', 'man' is the middle term). The *figure* of a syllogism is determined by the relative positions in premises and conclusion of the terms: in a first figure syllogism the middle term is on the left in the first (major) premise, on the right in the second (minor) premise, and the conclusion goes minor term, major term, as in the example above. The *mood* of a syllogism depends on the *type* of proposition making up the premises and conclusion. The types of proposition which occur are: universal affirmative propositions ('all A are B'); universal negative propositions ('no A are B'); particular affirmative propositions ('some A is B'); and particular negative propositions ('some A is not B'). Representing each of these types of proposition by the vowels *a, e, i, o*, respectively, the syllogism which I gave as an example above goes *aaa*, and this fundamental syllogism (the first mood of the first figure) was assigned the mnemonic *Barbara* (the vowels give the mood; the consonants also have mnemonic value, but I will not bother the reader with this here). The complete scheme of figures and moods of the syllogism was reduced to mnemonic form by Peter of Spain in the following verse:

Barbara Celarent Darii Ferio Baralipton
Celantes Dabitis Fapesmo Frisesomorum
Cesare Campestres Festino Baroco Darapti
Felapton Disamis Datisi Bocardo Ferison

(The first nine are first figure, but include the syllogisms later assigned to the fourth figure, the next four are second figure, the last six, third).

[3] *Enthymeme* is a contracted syllogism with one premise unstated, used particularly in informal argument to persuade rather than to prove. For example: 'It is lawful to repel force with force, therefore Clodius was lawfully killed by Milo.' Here the unstated minor premise is, 'Milo repelled the force of Clodius.' *Example* is a particular illustration of a general thesis, and was sometimes regarded as an enthymeme, sometimes as a contracted induction. For example: 'Abraham waged war for the safety of those nearest and dearest to him, therefore Christians may wage war to save their near and dear ones.' *Induction* establishes a universal proposition from a sequence of particular instances, and was regarded in the period as a legitimate form of inference if a complete enumeration is made of appropriate singulars, or if no counter example is to be found. The following sixteenth-century example of an induction is supposed to be an example of a perfect or valid induction: 'Adam, a blessed and pious man, had a cross to bear; So did Abel; So did

Spain's goal for dialectic is the formal, academic exercises of the schools. In contrast, Cicero and Quintilian, and the later 'humanist' dialecticians particularly favoured the weaker, persuasive types of argument as crucial tools in 'ordinary language' or oratorical discourse.

As supplementary aids for debating, Peter of Spain devotes tractates V and VI of his treatise to the *topics* or *places*, and the *sophisms* (the former based on Aristotle's *Topics* and Boethius' *De Differentiis Topicis*, the latter on Aristotle's *De Sophisticis Elenchis*). The topics provide a mnemonic classification of debating material for quick, ready retrieval in the formulation of a defence or opposition of a set theme. For instance (to use a sixteenth-century textbook example), suppose that the theme of a debate is the question, 'should a priest take a wife?'. All information which can be gathered together concerning 'priest' and 'wife' is assembled under a fixed list of headings.[1] This body of classified material is considered to provide all available arguments for or against the question. Under the heading *definition*, for 'priest' is included the information: 'desirous to live virtuously'. Under the same heading for 'wife' is included the information: 'received into the fellowship of life to avoid fornication'. From the place, *definition*, therefore, the following syllogism is constructed in support of the marriage of priests:

Whosoever desireth to live virtuously, desireth to avoid fornication
Whosoever desireth to avoid fornication and cannot obtain it by prayer or otherwise (as to all men it is not given) the same person desireth marriage
Ergo whosoever desireth to live virtuously [namely the priest] desireth marriage.[2]

The treatment of sophisms comprises an account of dishonest inferences based on ambiguity of terms, and invalid inferences with concealed false premises (such as Wilson in fact uses in the

Abraham; So did Jacob; So did Christ; nor are dissimilar cases to be found. Therefore all blessed and pious men have a cross to bear.' All these examples are taken from E. Sarcerius, *Dialectica* (Lipsiae, 1539); pp. 80, 88 and 85 respectively.

[1] The list of places which Cicero gives in his *Topics*, for cataloguing available information connected with the subject under debate, includes 'meaning', 'conjugates', 'genus', 'species', 'antecedents', 'contradictions'. This and a similar list in the *De Inventione* (I, xxiv–xxix) form the basis for the discussion of the places in the middle ages and renaissance. For a thorough treatment of Boethius' use of the topics or places see O. Bird, 'The tradition of logical topics: Aristotle to Ockham', *Journal of the History of Ideas*, 23 (1962), pp. 307–23.

[2] T. Wilson, *The Rule of Reason* (London, 1551), fo. 56r–58v.

example above). The sophisms supposedly trained the student to pick out fallacious arguments in an opponent's case, as well as protecting him against unconsciously introducing such arguments himself.

There is really no place for a treatment of the *Posterior Analytics* in this course. The *Summulae* instructs the student in the selection and manipulation of existing statements about the state of the world, and in the critical analysis of an opponent's arguments. It does not deal with fundamental questions about first principles, and it avoids philosophical issues which might raise such questions. Commentators on the *Summulae* who felt it necessary to include some treatment of Aristotle's treatise on demonstration for completeness acknowledged the limited aims of their textbooks, and confined themselves to a brief account of Aristotle's three simple criteria for testing that a proposition is 'scientific', that is, that it is an essential definition such that subsequent syllogistic inference which uses it as a premise will automatically yield a similarly 'scientific' conclusion. This preserves the impression carefully fostered throughout the *Summulae* that *truth* resides in language, and falsehood is detected by detecting faulty linguistic formulations of a case.

Tractate VII (the *parva logicalia*) completes the course with a relatively simple treatment of linguistic techniques developed in the middle ages for dealing with such problems as the denotation and connotation of terms, which make it possible to handle difficulties and detect flaws in arguments which the machinery of syllogistic analysis alone cannot make any impression upon.

I have set out the content of Peter of Spain's treatise in this amount of detail because we need to have before us a picture of the body of material which made up dialectic, and how it was presented to the student at the end of the fifteenth century. Peter of Spain's dialectic teaching rests on a meticulous classification of terms which enables the student to make a critical analysis of any proposition or sequence of propositions. Inevitably this means that a great deal of the student's time is taken up with terminological minutiae, although this terminology is, of course, supposed to be a shorthand for the philosophical and linguistic theories on which dialectical examination of problems is based. Thus the first chapter of Eckius' *Elementarius Dialectice* (1517), the elementary dialectic manual which he wrote to accompany his commented edition of the *Summulae* for students at the university of Ingolstadt,[1] requires that the student shall understand how to apply and distinguish categorematic and syncategorematic terms, absolute and connotative terms, signification and connotation, and terms of the first and second intention or imposition.[2] And although all these terms are historically important, and are related to particular mediaeval insights in logic, as far as the sixteenth-century student would have been concerned they are many of them redundant, and of no apparent use in the analysis of the types of argument he faced in the course of his university training for the Bachelorship (or even Mastership) of Arts.

It is from this narrow, pedagogic point of view that hostility towards the old logic (in particular the treatment of terms in the *logica vetus* and the *parva logicalia*) grew steadily towards the end of the fifteenth century. The alternative dialectic developed by Lorenzo Valla, Rudolph Agricola, Johannes Caesarius, and Philip Melanchthon (among others) owes much of its peculiar character to the particular requirements of teaching logic as part of a general arts course. Such reformers aimed to make the material of dialectic generally accessible, and to eliminate irrelevant and redundant material, rather than to summarise the whole mediaeval tradition in formal logic as preserved in an intricate and often obscure terminology.[3]

We may characterise the gradual modification of the dialectic handbook in the period between about 1450 and about 1550 as an attempt by educational reformers with a new view of the scope of higher education to break the hold of technical mediaeval logic on the university curriculum by reforms from the inside. Up to the end of the sixteenth century and beyond, in England at least, dialectic retained its central position in the curriculum of the four-year arts course for the first degree of BA; but by the 1530s the course was taught in conjunction with textbooks openly modelled on the pioneer reforming works of Lorenzo Valla, Rudolph Agricola and Philip Melanchthon, text-

[1] See Heath, art. cit., pp. 55–9.
[2] *Elementarius Dialectice* (Augustae Vindelicorum, 1517), fo. Aiir-Aiiiir.
[3] Heath has traced the way in which changes in the teaching of grammar which produced a move away from philosophical grammar and towards a descriptive grammar of Latin eloquence added to the inaccessibility of the terminology of scholastic dialectic to students by the early sixteenth century. See Heath, art. cit.

books which presented the subject in a very different light from Peter of Spain, or even his most enlightened commentators.[1] These reformers argued that dialectic belongs in the position assigned to it within the traditional classification of knowledge, at the focus of teaching and learning, because it embodies the principles and precepts vital to all intellectual processes. In this case it should, they argued, describe not simply how an artifical and academic subsection of language is deployed in formal disputation, but should provide an account applicable to all use of language, and all thought processes.

In practical terms this meant that the dialectician should set out his material in such a way as to make clear to the student how it relates to his everyday use of language, and should discard as irrelevant needless detail accumulated over the centuries in the course of close study of the standard logic texts. In attempting to do this, the reformers with whom we are concerned were led to emphasise two points in particular as central to their view of dialectic. In the first place they insisted that Aristotle's distinction between correct logical argument from plausible or probable premises (dialectic) and correct logical argument from certain and incontrovertible premises (demonstration) was for their purposes a meaningless one, since the same techniques are involved in argument of either type. They identified dialectic, the *ars disserendi*, with the whole of logic as the general field of rational argument embracing probable, certain and sophistical argument, on the grounds that study of techniques of argument does not depend on the status, whether certain, probable or dubious, of the material to which these are applied.

This view is entirely consistent with their general (and essentially *humanist*) concern with the display and presentation of an existing body of knowledge; it is at the aspiring lawyer, teacher, cleric or politician that their course of instruction is primarily aimed. In the second place, therefore, they maintained that dialectic's primary field of application is all those areas of ex-

[1] On the place of dialectic teaching in Cambridge in the sixteenth century see L. A. Jardine, 'The place of dialectic teaching in sixteenth century Cambridge', *Studies in the Renaissance* (1974). On its place in the mediaeval curriculum see J. A. Weisheipl, 'Curriculum of the faculty of arts at Oxford in the early fourteenth century', *Mediaeval Studies* 26 (1964), pp. 143–85, and 'The place of the liberal arts in the university curriculum during the XIVth and XVth centuries', *Actes du Quatrième Congrès International de Philosophie Médiévale* (Montreal, 1969), pp. 209–13; M. H. Curtis, *Oxford and Cambridge in Transition 1558–1642* (Oxford, 1959), pp. 83–125.

change of knowledge which can loosely be grouped together under the heading of 'teaching'. This covers all communication of a body of knowledge by an expert to a novice, and includes both didactic instruction and persuasive oratory. To fit the study of dialectic to this approach the innovators adjusted its definition (as Cicero had done before them) to bring it into line with definitions of oratory or *rhetoric*. As we shall see, the outcome of these two crucial modifications was a dialectic which was supposed to be at once able to account for all but the most ornamental techniques of oratory, and to be suitable as a vehicle for both rigorous and informal argument, and in particular for teaching.

Because of the breadth of application claimed for this dialectic, the ambiguity of the classical accounts of the power of dialectic assumed considerable importance. For Plato, dialectic is definitely an instrument for discovery of truth.[1] Aristotle's position is less clear cut. Sometimes he contrasts dialectic with demonstration; that is, he contrasts plausible arguments based on probable premises with scientific inference based on certain premises.[2] In this context he maintains that even demonstration cannot establish first principles, but that these must be taken on trust from the theory of the science concerned, at the outset of debate.[3] First principles are arrived at by an induction or ἐπαγωγή.[4] But Aristotle also characterises dialectic as a question and answer dialogue, and this characterisation appears consciously to include Platonic dialectic. Aristotle does say in several places that by a critical question and answer examination of accepted general beliefs or ἔνδοξα which are the starting point of dialectic it is possible to arrive at fundamental principles, that is, to discover truths unknown at the outset (and not contained in the premises of the argument).[5] And this appears to be the procedure he uses in the *Physics*, *Metaphysics* and *Nicomachean Ethics*.[6]

[1] For a clear, brief summary of Platonic dialectic see H. D. P. Lee, 'Geometrical method and Aristotle's account of first principles', *Classical Quarterly*, 29 (1935), pp. 113–24, p. 119.

[2] *Posterior Analytics* I, 1 (71a1-71b7); *Topics* I, 1 (100a25-101a25); *Metaphysics* IV, 2 (1004b15-30).

[3] *Topics* I, 1 (100b22).

[4] *Posterior Analytics* II, 19. For a useful general account of ristotle's views on induction see G. E. R. Lloyd, *Aristotle: the Growth and Structure of his ιought* (Cambridge, 1965), pp. 125–7.

[5] *Topics* I, 2 (101a37ff.).

[6] See Lee, art. cit., pp. 122–3; G. E. L. Owen, 'Tithenai ta Phainomena', *Aristotle: A Collection of Critical Essays*, ed. Moravcsik (London, 1968), pp. 167–90.

Some of the dialectical reformers restricted themselves to discussion of teaching and presentation, on the grounds that this was the essential area covered by the *ars disserendi*. The techniques of dialectical presentation, including the procedure of clear and comprehensive presentation of an extended argument or narrative, or of an entire discipline, which came to be known as dialectical '*method*' or '*ordo*', provide precisely those skills (clear explication and precise itemisation of content) which the teacher habitually uses. And amongst these teachers are included lawyers and professional men who rely on the same techniques. But there were other dialecticians who chose to underline the links with Platonic dialectic and the Platonic strains in Aristotle.[1] For such authors there is at least the suggestion that dialectic can corroborate the material selected and presented. This outlook goes along with claims that the techniques of dialectic must be those of 'natural reason', and dialectical method must mirror our apprehension and classification of natural phenomena. The claim that dialectic is the universal vehicle for teaching (as made, for instance, by Agricola) became for such dialecticians a claim that dialectic is an arbiter of truth.

In the discussions of *method* with which for our present purpose we are particularly concerned, this ambiguity in the status of dialectical presentation caused considerable confusion. There are, in classical and renaissance writings on the subject, three main, distinct contexts for discussion of an extended ordering of a sequence of propositions – what the renaissance called *method*.[2] Firstly there is a discussion of axiomatic method, ultimately based on Euclid's *Elements* and Proclus' commentary on Euclid (first translated into Latin in 1560).[3] Secondly there is the

[1] It is in fact noticeable that most of the reforming dialecticians I shall be discussing draw Platonic dialectic into their introductory discussion of the scope of dialectic.

[2] The broad definition of the term as the renaissance understood it is well represented by Melanchthon's general definition in his *Erotemata Dialectices* (1547): 'The Greeks defined it thus: ...Method is an acquired aptitude, that is a science or art, for making a way according to a certain plan; that is, method discovers and opens a way as if through places blocked and made impassible by thorns, and produces material pertinent to a theme, and sets it out in an orderly fashion' (Wittebergae, 1555 ed., p. 104). 'Sic Graeci definiunt: ...Methodus est habitus, videlicet, scientia, seu ars, viam faciens certa ratione, id est, quae quasi per loca invia et obsita sentibus, per rerum confusionem, viam invenit et aperit, ac res ad propositum pertinentes, eruit ac ordine promit.' For a general survey of method in the renaissance (which unfortunately does not always separate out the various traditions involved) see N. W. Gilbert, *Renaissance Concepts of Method* (New York, 1960).

[3] The first printed Greek text was published in Basel in 1533. The Latin translation was by the Paduan F. Barozzi. See G. R. Morrow, *Proclus, A commentary on the First Book of Euclid's Elements* (Princeton, 1970), p. xliv.

long-standing discussion of methods of demonstration, primarily based on Aristotle's *Posterior Analytics*, but often referring to Plato's theory of dialectic, Galen's remarks on method in the *Ars Parva* and Averroës' remarks in his commentary on Aristotle's *Physics*.[1] Thirdly there is the discussion of methods for laying out available material for clarity, sometimes distinguished under the alternative term *ordo*, and based on sources which include those for the discussion of methods of demonstration, together with remarks in Cicero, Quintilian, and in the pseudo-Cicero's *Ad Herennium*.[2] As we shall see, there is a marked tendency amongst the reforming dialecticians to confuse these contexts in their discussions of method. Indeed, this is one of the bitterest criticisms levelled against them by competent orthodox Aristotelians (like Scheckius and Zabarella), who were perfectly capable of keeping the contexts apart, and distinguishing the different intended functions of the various types of ordering procedure. As we shall also see, awareness of latent confusion within the dialectical tradition lies at the root of Bacon's dissatisfaction with sixteenth-century reformed dialectic, and shapes his own attempts to formulate an adequate method of discovery.

Dialectical methods

Rudolph Agricola's *De Inventione Dialectica libri tres* provides the natural starting point for an account of the reforming of dialectic teaching and the subsequent emergence of the topic of dialectical method. First published in 1515, it continued to influence writers on dialectic until the turn of the century.[3] In Cambridge, the 1535 university statutes specify Agricola's handbook as the model for dialectic teaching for arts students, the 1560 Trinity College statutes give it as one of the texts from

[1] References for these sources are given where they are discussed in more detail below.

[2] Baldwin and Zabarella, amongst the authors I have consulted, refer to these methods as *ordines*, and distinguish them clearly from methods of demonstration (called by Zabarella *methodi*). Bacon restricts the term *methodus* to methods of presentation, and uses the terms *via* and *ratio* for methods of discovery (Cicero renders the Greek μέθοδος 'via ac ratio'). Amongst recent authors who have tackled the subject of method in the renaissance only Schüling attempts to sort out the separate contexts for the various discussions. See H. Schüling, *Die Geschichte der axiomatischen Methode im 16. und beginnenden 17. Jahrhundert* (Hildesheim, 1963).

[3] Rodolphus Agricola (1444–85). For biographical details see C. Vasoli, *La dialettica e la retorica dell' Umanesimo* (Milan, 1968), pp. 147–82. The *De Inventione* was finished in 1480, but first published in 1515, according to A. Crescini, *Le origini del metodo analitico: il cinquecento* (Udine, 1965).

which college lecturers are to teach, and a remarkable number of the booklists amongst the Cambridge probate records for the period 1540–90 include a copy of this work.[1]

Peter of Spain had defined dialectic as 'the art of arts and science of sciences, giving access to the principles of all disciplines',[2] and on etymological grounds he had characterised it as an art and science of disputation. In contrast, Agricola's definition of dialectic at the opening of the second book of his *De Inventione* specifies a teaching context for dialectic:

> We shall agree to define dialectic as the art of discoursing plausibly concerning any set theme, as far as shall be possible in the circumstances... This then shall be the goal of dialectic, to teach according to the nature of the thing about which one is discoursing, that is, to discover those things which are apt for establishing belief, and to lay out what has been discovered, and to order them as is best suited for the purpose of teaching.[3]

Agricola deliberately confines himself to discussion of selection and display of material for clear and concise teaching, with the proviso that all clear and accessible discourse may be deemed to be teaching something to someone. If dialectic is an art, Agricola argues, as the traditional definition requires, then it must by definition contain a body of precepts useful for life in some field.[4] The precepts of dialectic undoubtedly refer to discourse, and the practical field of application of discourse is teaching; hence a teaching context should be specified at the outset of discussion of dialectic. Like Valla before him (in his *Dialecticae Disputationes*

[1] For the University statutes see *Documents Relating to the University and Colleges of Cambridge* (London, 1852); for the Trinity statutes see J. B. Mullinger, *The University of Cambridge* (Cambridge, 1884) (appendix to vol. II); the student booklists are in the Cambridge University Archives.

[2] 'Dialectica est ars artium scientia scientiarum ad omnium methodorum principia viam habens. Et ideo in acquisitione omnium scientiarum dyalectica dicitur esse prior. Dicitur enim dyalectica a dya quod est duo et logos quod est sermo vel lexis quod est ratio / quasi duorum sermo vel ratio scilicet opponentis et respondentis in disputando.' *Expositio magistri Petri Tatareti in summulas Petri Hyspani* (Limoges, *c.* 1510), fo. Aiiv. The ultimate source for such definitions is Aristotle, *Topics* I, 1.

[3] 'Erit ergo nobis hoc pacto definita dialectice, ars probabiliter de qualibet re proposita disserendi, prout cuiusque natura capax esse poterit...Hic itaque finis erit dialectices, docere pro facultate rei de qua disseritur, id est invenire quae fidei faciendae sint apta, et inventa disponere, atque ut ad docendum quam accommodatissima sint ordinare', *De Inventione* ([Colonie], 1528), pp. 155 and 158. This definition depends on Cicero's formulation of invention and judgment in *Topics* II, 6. Indeed, Agricola's entire treatment of dialectic owes much to Cicero.

[4] For Agricola's argument that dialectic must give a set of practical precepts if it is to be an art see *De Inventione*, ed. cit., pp. 152–3.

of 1439), Agricola makes a particular point of the fact that Peter of Spain's etymological definition is incorrect – dialectic is derived simply from the Greek, διαλέγομαι, 'I discourse' – and that in consequence he has confined his study to a narrow and artificial branch of language usage in keeping with his narrow and artifical definition.

Peter of Spain's characterisation of dialectical argument included the term '*probabiliter*', which was understood (following Aristotle's distinction between probable and apodeictic reasoning in *Topics* I. 1) to cover statements which are 'plausible', 'believed by the majority', or 'which hold in most cases', and which are therefore suitable as the premises for academic disputation. In this context it is more important that the material shall be readily acceptable, and not particularly vulnerable to attack, than that it shall be incontrovertibly and absolutely true. Agricola retains the term, but explicitly rejects the view that it implies any uncertainty about the overall truth or falsity of dialectical argument. For him 'probabiliter' is to be taken as including whatever is 'apt' or 'appropriate'; whatever is fitting to a discourse may be spoken '*probabiliter*'.[1] Dialectic is thus taken as including in its applications discourse which employs such 'apt' devices as fable and parable, and hence the range of Agricola's 'teaching' extends to include all narrative literature.

Given this deliberate shift of focus away from metaphysics and formal disputation on technical topics, towards the literary, it is not surprising that Agricola models his work much more closely on Cicero and Quintilian than on Aristotle. (Valla, to whom Agricola refers with admiration, goes to considerable lengths in his handbook to show that the knotty metaphysical problems of scholastic logic are in fact questions of grammar and language usage; for him Quintilian's treatment of dialectical topics is the model for an account correctly balanced between technical detail and the requirements of the practising orator or teacher.) One of the marks of this preference for the Latin dialecticians is Agricola's division of dialectic into two parts, corresponding to two distinct stages in the process of formulation of discourse,

[1] 'Id nunc dicamus, quia volumus dialectices esse, posse de quolibet dicere probabiliter: probabile in disserendo non solum id esse, quod revera probabile est, hoc est, quemadmodum Aristoteles inquit, quod vel omnibus videtur, vel plurimis, vel sapientibus ...nobis erit probabile, quod apte consentaneeque de re proposita dicetur.' ed. cit., pp. 154–5.

inventio and *judicio*.[1] Invention is the process of selection and collation of material; judgment its organisation for presentation. In terms of the traditional arrangement of the Aristotelian corpus, invention makes use only of the places or topics; judgment incorporates the treatment of the predicables, predicaments, types of proposition and types of inference. Like Cicero in his work of the same name, Agricola elects to concentrate on the treatment of invention, or the places, which he regards as an important and neglected area of the dialectic course.

Agricola's emphasis on invention, like Cicero's before him, signals an approach to dialectic which aims at extracting from Aristotelian logic the bare minimum of formal apparatus, and building on this a largely descriptive study of language usage, applicable in any area of discourse. The classificatory scheme of the places of invention had traditionally featured both within the dialectic and the rhetoric course, in barely differing forms, as the general procedure for assembling and classifying the available material on a set theme. By making this the focus of his discussion of dialectic Agricola is enabled to combine in a single *ars disserendi* material traditionally separated under the headings 'rhetoric' and 'dialectic'. The first book of his treatise gives a detailed account of the places[2] in such a way as to support the view that they suffice to provide a comprehensive classification of all knowledge. This is followed, in book two, by an extended discussion of the field of application of dialectic (*quaestio*, or any general or particular question whose answer requires deliberation), and its tool or instrument, articulate speech (*oratio*). The discussion of *oratio* enables Agricola to incorporate material conventionally assigned to rhetoric, including *status*,[3] *copia*,[4] the

[1] Cicero maintains that 'omnis ratio...disserendi' falls into two parts: invention of arguments according to a list of places adapted from Aristotle's topics, and the use of these places for judging the validity of some 'quaestio' (*Topics* II, 6; XXI, 79). Invention according to the places involves cataloguing the available information connected with the subject under debate according to its relation to that subject.

[2] For a comprehensive table of the places of Cicero, Themistius and Agricola, see Agricola, ed. cit., p. 136. Vasoli suggests that Agricola's source for the places as given by Themistius was Boethius (op. cit., p. 174). For Agricola's own account of his sources see Agricola ed. cit., pp. 14–15, and for a summary of the places see pp. 21–2.

[3] *Status*, as discussed by Cicero in *Topics* XXV, and under the term '*constitutio*' in *De Inventione* II, is the general type of question under which a particular argument may fall, and according to which the overall strategy for arguing the case is decided. See also Quintilian, *Institutio Oratoria* III, VI.

[4] *Copia* embraces the discussion of all devices for amplifying, embellishing and illustrating discourse. The topic is established as an important part of the study of eloquence by Erasmus in his *De duplici copia verborum ac rerum commentarii duo* (Parisiis, 1512).

parts of an oration (the elementary divisions into which a speech falls: introduction, narration, arguments pro and con the theme, conclusion), decorum, style and (in book three) procedures for moving the affections of an audience. These topics derive ultimately from Cicero, Quintilian, the *Ad Herennium*, and Aristotle's *Rhetoric*.[1] The ease with which Agricola is able to introduce this material into a natural account of dialectic supports his contention that a single study of the art of discourse, dialectic, provides all that is necessary in practice in fields of discourse traditionally separated into the dialectical and the rhetorical. Embellishment and ornamentation, and techniques of actual delivery of orations are the only aspects of discourse which Agricola is prepared to regard as a separate study (rhetoric).

At the end of the *De Inventione* Agricola includes a treatment of what he calls '*dispositio*': the ordering of the material of a discourse. He takes as the basis for his discussion a division of disposition into natural, arbitrary and artifical.[2] Natural order follows the order of things in nature, and is of four kinds: temporal order, order of existence (from genus to species, parts to whole, causes to effects or outcomes, subject to adjunct), geographical order (west to east, say, or from head to feet) and order of social status. These are the 'orders of priority' traditionally introduced into the discussion of the 'post-predicaments'. Arbitrary order arranges things for which there is no obvious natural ordering in any order the author chooses. Artificial order reverses the natural ordering for dramatic or other effect.[3] Agricola's discussion is more comprehensive than that of the

[1] Cicero and Quintilian are concerned first and foremost with the study of law and techniques of legal argument. These are both argumentative and rhetorical, and hence their works were apparently seen as bridging rhetoric and dialectic. For a general account of Greek rhetoric, and in particular the works of Hermagoras, on which Cicero and Quintilian's treatments are based, see G. A. Kennedy, *The Art of Persuasion in Greece* (London, 1963), especially pp. 303–18. See also D. L. Clark, *Rhetoric in Greco-Roman education* (New York, 1957). On the important question of *status* or *stasis*, as introduced by Hermagoras, see Kennedy, op. cit., pp. 306–14; and for further interesting discussion of Cicero's use of *status* see W. S. Howell, *The Rhetoric of Alcuin and Charlemagne* (Princeton, 1941).

[2] This division has affinities with divisions of disposition in the traditional rhetoric manual, based on the *Ad Herennium* passage: 'Genera dispositionum sunt duo: unum ab institutione artis profectum, alterum ad casum temporis adcommodatum' (III, ix, 16). Natural ordering follows the rules of rhetoric, artificial follows the order of the occasion. See for instance G. Trapezuntius, *Rhetoricorum libri V* (Venetiis, 1523), fo. 53r. On the importance of Trapezuntius' rhetoric handbook see R. D. Gooder, *The Rhetorical Work of Juan Luis Vives* (unpublished Ph.D. diss., Cambridge, 1969), chapter IV, 'Rhetoric and Style in the Renaissance', pp. 114–28.

[3] Agricola, ed. cit., pp. 357–8.

conventional rhetoric handbook (which traditionally contained a brief account of disposition). It includes syllogism and induction as natural orderings of *argumentatio*,[1] and it singles out as particularly suitable for teaching that order which proceeds from the most general precepts to the less general.[2]

Agricola's teaching method which moves from the most general principles to the less general is based on Aristotle's stipulation that in the pursuit of knowledge one should proceed from what is better known intellectually (the most general principles) to what is less well known. Aristotle recognised two senses in which the term 'prior' or 'better known' may be used. Something is 'better known to us' the more readily it is perceptible to our senses. On the other hand, the more general a proposition, the 'better known in nature', that is, in itself, or intellectually. Scientific or certain knowledge, according to Aristotle, must begin with what is better known in itself.[3] In *Topics* VI, 4, however, Aristotle stated that although the ideal way of conveying knowledge of a subject is to commence with the most general definition (that which is prior in nature), it may be necessary when instructing a novice to begin with what is better known to him, that is, with some less general proposition, and proceed to the most general. Thus whilst 'point' is the most general notion in geometry (and Euclid's *Elements* opens with its definition), the student may be more familiar with lines and surfaces (which Euclid defines after 'point'), and it may prove easier for him if 'point' is defined in terms of these (141b2-15). Agricola, however, takes it that the most general propositions are, for teaching purposes, unequivocally 'better known'.[4]

Agricola does not propose a general theory of ordering. He acknowledges that this is an area which resists generalisation, and which has been neglected in classical treatments of dialectic

[1] Agricola, ed. cit., p. 388.

[2] 'A generalioribus, quoniam notiora sunt, semper ducendus erit ordo. Deinde per species eundum. Et hae ipsae species rursus fortasse, si tam late pateant, genera fient, inque alias erunt species diducendae. Singula ergo, ut primum quodque sumimus, sic erunt deinde tractanda: ut dicatur, quae sit substantia eorum, quod definiendo fit: si tamen eiusmodo res sit, ut sit definienda: notiores enim sunt nonnunquam, quam ut hac indigeant cura.' Agricola, ed. cit., p. 363.

[3] *Posterior Analytics* I, 2 (71b27-72a6); see also *Metaphysics* I, 2 (982a20-25).

[4] There appears to be some difference of opinion in the renaissance as to whether the phrase for intellectual priority is 'notiora natura' or 'notiora naturae', and whether the phrase is to be interpreted 'better known *in* nature' or 'better known *to/for* nature'. See L. A. Kosman, *The Aristotelian Backgrounds of Bacon's 'Novum Organum'* (unpublished Ph.D. diss., Harvard, 1964).

and rhetoric.[1] But he makes an attempt to gather together in one place the scattered remarks on presentation of complete discourses to be found in Aristotle, Cicero and Quintilian. His discussion establishes an important precedent for sixteenth-century dialecticians. By discarding all material which is not directly relevant to the teacher (and a teacher whose aim is clarity and eloquence in his own presentation, and the acquisition of eloquence by his students) he produces a handbook whose new, clear focus has radically changed its character as a teaching aid. And this is achieved not simply by discarding unnecessary detail and sophisticated treatment of logical minutiae, but also by elaborating on passing remarks in existing treatments to create new topics.

Under Agricola's influence, Philip Melanchthon defined dialectic as follows in his *Erotemata Dialectices* (1547):[2]

> Dialectic is the art or way of teaching correctly, perspicuously and in an orderly fashion, which is achieved by correctly defining, dividing, linking true statements, and unravelling and refuting inconsistent or false ones.[3]

Here the teaching emphasis of Agricola's definition is retained, and hence the innovatory outlook towards dialectic teaching of Agricola's manual. But by reinstating judgment, and with it most of the traditional treatment of terms and propositions, Melanchthon produces a textbook less drastically unconventional in format than Agricola's. In practical terms this made Melanchthon's handbooks, like Johannes Caesarius' *Dialectica*[4]

[1] Agricola, ed. cit., p. 356.

[2] Philip Melanchthon (1497–1560). For biographical details see Vasoli, op. cit., pp. 278–309.

[3] *Erotemata Dialectices, continentia fere integram artem, ita scripta, ut iuventuti utiliter proponi possint* (1555 ed.), p. 1. 'Dialectica est ars seu via, recte, ordine, et perspicue docendi, quod fit recte definiendo, dividendo, argumenta vera connectendo, et male cohaerentia seu falsa retexendo et refutando.' In his earlier *Dialectices libri quattuor* (1527) Melanchthon is more specific, though less concise: 'Dialectica est ars ac via docendi. haec enim est proprie vis dialecticae. Porro omnis docendi via ac ratio consistit in definiendo, dividendo, et argumentando. Definimus autem, cum aut nomen interpretamur, aut quid res sit exponimus. Dividimus, cum membra aut partes enumeramus. Argumentamur, cum aliam sententiam ex alia ratiocinamur...Est autem proprium et principale officium dialectices, docere. Id fit definiendo, dividendo, ratiocinando, quae quam afferant utilitatem, postea cognoscetur ex praeceptis.' cited from *Opera* (Basileae, 1541), vol. V, p. 172.

[4] Joannes Caesarius (1467–1550). For biographical details see Vasoli, op. cit., p. 260. For the widespread influence of Caesarius' *Jo. Caesarii viri undecumque doctissimi Dialectica* (1532) see Risse, op. cit., pp. 25–31. For the extremely wide dissemination of Melanchthon's work see Risse, pp. 79–121; Vasoli, p. 278ff. Caesarius, like Melanchthon, acknowledges his debt to Agricola, and he adopts Agricola's definition of dialectic. See *Dialectica* (1568 ed.), fo. †8r.

(which is in very much the same spirit, and adopts the same approach) more suitable as a means of introducing students to the standard corpus of Aristotelian works and their commentaries than the *De Inventione Dialectica*. Nevertheless, Caesarius and Melanchthon do succeed in preserving the flavour of Agricola's work. Dialectic is presented as a general rather than a specialist study, to be applied wherever the rules of natural discourse and reason are appropriate. And because in following the traditional order of the dialectic curriculum (the order of Peter of Spain's *Summulae*) both authors are obliged to treat judgment before invention (since the topics or places are treated *after* terms and propositions and the syllogism in this scheme), the conflict between the traditional 'Aristotelian' approach and the 'Ciceronian' approach of Agricola is forced out into the open. In support of the traditional ordering, both authors produce the argument that although invention certainly precedes judgment in our actual formulation of discourse, in the order of clear teaching judgment needs to be explained before invention.[1] Such an argument is entirely typical of the new approach; philosophical considerations are displaced by practical questions of expediency.

Seton's little elementary handbook, the *Dialectica* (1545), which may have been an elementary set text in dialectic at Trinity (in conjunction with Agricola) when Bacon was a student there, is a particularly good example of the type of dialectic I am here concerned to characterise. Seton states in his preface that he aims to give a simple treatment of the two texts specified in Henry VIII's university statutes, Agricola and Melanchthon.[2] Like Melanchthon and Caesarius, he uses the traditional format for his handbook; the first three books handle judgment, whilst the fourth handles invention (and is directly based on Agricola's account of the places of invention).

Seton's first two books treat schematically, and in much the traditional order, terms and propositions (the predicables, predicaments, types of proposition, conversion, etc.). There is no

[1] For the division of dialectic into invention and judgment see Melanchthon, *Erot. Dial.* ed. cit., pp. 224–5; Caesarius, *Dial.* ed. cit., fo. M5v-M6r. For a typical justification of the reversal of these in presentation see Raiianus, commenting on Caesarius' *Dialectica*, ed. cit., fo. M7r: 'Inventio ordine naturae prior est iudicio (prius enim invenire oportet, deinde iudicare de inventis.) Iudicium autem ordine doctrinae, commoditatis, facilitatis, est prius inventione.'

[2] See Seton, *Dialectica* (Londini, 1584), fo. *A*2r.

suggestion that these provide more than convenient means of sifting and classifying material, whether for analysing an existing text, or as a preparation for debate. The treatment culminates, at the beginning of Book 3, in a short essay on *disputation*, and the various strategies to be adopted in the defence or opposition of a set theme. Thus although we have the illusion of having run over the Aristotelian set texts in order, the suppression of the philosophical and speculative questions (particularly evident in the extended commentaries which always accompany the Peter of Spain text) effectively reduces the treatment to that of Aristotle's *Topics* alone (the practical treatise on the use of logic for analysis and construction of discourse in debate).

This is particularly apparent in Book 2, which includes brief sections on *definition* and *division*, which were traditionally assigned an important rôle in the Aristotelian logical system.[1] In his treatment of these topics Seton shows himself in two ways to belong to the new school of thought in dialectic. First, he passes over the rôle of definition and division in the establishing of scientific principles, and presents them as part of the critical apparatus for analysing propositions and building supporting arguments in disputation – the rôle assigned to definition and division in the *Topics*. Secondly, Seton is at pains to make his accounts broad enough to include both the formal use of definition and division as analytic tools, and the less rigorous associated procedures for literary or 'natural' discourse. Alongside definition by essential characters he discusses *descriptio* (rhetorical definition by description of visible and inessential characteristics), definition by etymology and definition by metaphor.[2] Alongside division of a genus into its fixed subspecies he discusses *partitio* (sequential description of, for instance, visible parts of a whole, for display or clarity of expression).[3] In this way he widens the area of application of his dialectical rules to include most of literature, and not simply formal disputation.

It does not therefore come as a surprise that only a comparatively short section of Book 3, which handles *argumentatio*, is devoted to the figures of the syllogism, which formed the pivot of the old dialectic course. In fact, Seton's treatment amounts to little more than a checklist of rules to be observed in framing

[1] See the next section on methods of demonstration.

[2] *Dialectica*, ed. cit., fo. I vir-Kir. [3] *Dialectica*, ed. cit., fo. Kir-Kivr.

syllogisms, and in inspecting the syllogisms of others. On the other hand, he discusses less rigorous and conventional argument forms, which Peter of Spain had all but ignored. In addition to enthymeme, induction and example, which did traditionally receive some mention alongside the syllogism, he includes discussion of *dilemma* ('if you marry a comely wife you will share her, if you marry an ugly one it will be irksome to you, accordingly you should marry none'[1]), *sorites* ('man is an animal, an animal is a living thing, a living thing is a substance, therefore man is a substance'[2]), and what he calls '*rhetorical syllogism*'. This last is any extended, discursive argument (Seton's example is borrowed from Cicero's *De Inventione*) which may be broken down roughly into 'statement' (*propositio*), 'procedure' (*ratio*), 'assumption' (*assumptio*), 'support of the assumption' (*assumptionis probatio*), and 'consequence or outcome' (*complexio*).[3] Throughout his discussion Seton stresses the flexibility of the various forms (both in his text and in his examples), and not the comparative inferential rigour or '*vis illatrix*', in which Italian dialecticians were, in the same period, chiefly interested.[4] The important impression which one carries away from Seton's textbook is that it is designed to serve as well as a guide to analysing a Cicero oration or a passage from Ovid as for formal debating.

Melanchthon, Caesarius and Seton do not any of them include a treatment of the general ordering of discourse which Agricola called '*dispositio*'. However, Melanchthon does discuss the question of a general procedure for organising the available material in order to investigate a 'simple theme' (a simple theme is a question like 'what is virtue?'; a compound *quaestio* or theme is a question like 'pleasure is not the true end of man',[5] which according to Melanchthon is investigated by breaking it down into component simple themes). Adopting a term with wide connotations from classical Greek Melanchthon chose to call this procedure 'method'.[6] His technique for handling simple themes

[1] *Dialectica*, ed. cit., fo. Ovir.
[2] *Dialectica*, ed. cit., fo. Nvir.
[3] *Dialectica*, ed. cit., fo. Nvr.
[4] See next section on methods of demonstration.
[5] 'Voluptas non est finis hominis', Melanchthon, *De Dialectica* (Witebergae, 1531 ed.), fo. L2v.
[6] In the *De Dialectica* the section on the handling of a simple theme is headed 'de modo explicandi simplicia themata, et quomodo sit utendum his praeceptis, quae hactenus tradita sunt' (1531 ed., fo. E5r). When, however, Melanchthon discusses *quaestiones* in the fourth book, in the context of general dialectical procedure, he refers to this earlier

combines the four *quaestiones* of the *Posterior Analytics* which preface all investigations of phenomena, with extra questions based on the places, to give an ordered list of headings under which the material of any investigation ought to be considered; what the word (*vocabulum*) signifies; if it exists; what it is; what its parts are; what its species; what its causes; what its effects; what things agree with it; what things are in opposition to it (*pugnantia*).[1] Melanchthon stresses the need in all presentation for clear and orderly display of the stages of the argument, and he offers 'method' as the key to such clarity. Method, in fact, for Melanchthon, is the way in which one thinks out the stages in the treatment of a question, and then sets out those stages as the most perspicuous presentation of the material.[2] And this means that there is a measure of ambiguity in Melanchthon's account of method; it is not in fact clear whether he is advocating a procedure for *deriving* or *presenting* material. As we shall see, in discussing methods of demonstration, Melanchthon consistently blurs the distinction between these two processes.

Melanchthon's brief remarks on method were taken up by two dialecticians in particular as offering a general account of the strategy to be followed in all extended argument or presentation. Jodocus Willichius and Nicolaus Hemmingius both wrote extended treatises on 'method', in which they attempted to put the dialectical discussions of ordering on an equal footing with existing accounts in the context of geometry and demonstration.[3] They added to Melanchthon's 'method for handling a simple theme', three methods for handling a composite theme, derived

discussion as a treatment of 'method': 'Ad hunc modum duci per singulas quaestiones res proposita debet, quod cum sit, et nihil effugiet, vel docentes, vel discentes, quod ad institutum pertinet...Nec perfecte putet se quispiam rem tenere ullam, nisi aut per omnes huius artis locos, aut per praecipuos ierit, et inquisiverit, ea, quae iubet unusquisque locus inquiri. Quare semper in conspectu nobis hi loci debent esse, ut quocies aliquid inciderit, statim percurrat eos ordine animus, et quaerat, quid dici conveniat...Veteres Methodum vocant rationem recte atque ordine docendi, iuxta praecepta dialectices, ac saepe monent ut in omnibus negociis, controversiis, artibus, demus operam, ut Methodum teneamus, quia necesse sit animum vagari incertum, nisi hac ratione regatur.' ed. cit., fo. L3r-v. In the *Erotemata Dialectices* the section on handling a simple theme is entitled 'De Methodo' (1555 ed.), p. 104.

[1] Aristotle's discussion is in *Posterior Analytics* II, 1. In the *De Dialectica* Melanchthon gives only Aristotle's four questions as the basis for handling a simple theme. In the *Erotemata Dialectices* they are expanded to the list above: 'Aristoteles in libro 2. Resolutionum posteriorum quaestiones recenset quatuor...Nos paucas addidimus' (1555 ed.), p. 105.

[2] *De Dialectica* (1531 ed.), fo. L3v-L4v; *Erotemata Dialectices* (1555 ed.), p. 227.

[3] Jodocus Willichius (1501–*c.* 1555); Nicolaus Hemmingius (1513–1600). See Willichius, *De Methodo omnium artium et disciplinarum* (Francof. ad Viadrum, 1550); Hemmingius, *De methodis libri duo* (Wittebergae, 1559).

from Galen: analysis (or resolution), synthesis (or composition), and definition (diaeresis or horistica).[1] The source for these methods is the highly ambiguous passage which opens Galen's *Ars Parva*.[2] Renaissance authors expanded these by collating them with other passages in Galen's works, and in Averroës, and their interpretation varies widely according to their particular selection of sources. Some authors related Galen's methods of analysis and synthesis to the corresponding methods of geometrical demonstration; others equated them with syllogistic arguments from effect to cause and cause to effect; and yet others with methods of presentation, the one proceeding from the complex and particular to the simple and general and the other proceeding conversely.[3] The third of these methods, definition or diaeresis, was generally identified with Plato's method of definition by successive dichotomy. Hence although Galen's methods are, in context, clearly intended as ways for laying out material for teaching, some renaissance writers tried to understand them as methods of discovery or demonstration, perhaps because of the particular misleading possibilities inherent in Galen's subject matter, medicine.[4] A method for laying out symptoms of diseases in order to teach diagnosis may be read by an undiscerning reader as a method for *discovering* the cause of the symptoms. By invoking the Galenic methodological tradition dialecticians like Hemmingius and Willichius laid methods of teaching and presentation of what is already known to the teacher open to confusion with methods for deriving unknown principles. The dialectician who made the most of the ambiguities inherent in

[1] See, for example, Willichius, ed. cit., fo. 2r: 'Methodus duplex est, aut simplex aut composita. Illa quidem est dispositionis artificiosae in docendi et discendi via, atque circa simplicia themata versatur. Haec vero est inventionis et probationis et postea compositi thematis. Sicut haec ad confirmationem rei et confutationem conducit, ita illa ad ordinatam quandam recteque dispositam explicationem confert. Sunt autem tres, neque plures esse possunt. Cum enim ordine a certa ratione tradere aliquam artem voluerimus...vel a fine, vel a principio, vel a totius comprehensione initium apte sumi ad artificiosum et legitimum progressum potest. Igitur necessario tot erunt rationes docendi seu methodi, una est ἀναλυτική, altera συνθετικὴ...ultima est ὁριστική...Latinis sunt resolutoria, compositiva et definitiva.' cit. Risse p. 110.

[2] See Galen, *Opera Omnia*, ed. Kuhn (Lipsiae, 1821), vol. I, p. 305. I give Kuhn's Latin translation for convenience: 'Tres sunt omnes doctrinae, quae ordine comparantur. Prima quidem ex finis notione, quae per resolutionem fit; secunda ex compositione eorum, quae per resolutionem fuerunt inventa; tertia ex definitionis dissolutione, quam nunc instituimus.'

[3] See appendix for a note on the source of ambiguity of these terms.

[4] Melanchthon confuses the Galenic teaching methods with geometric methods of axiomatic proof. See below p. 50.

the discussions of method in dialectic, and of dialectic as a whole in the period, was Peter Ramus.

Ramus defined dialectic succinctly as 'the art of discoursing well, and in the same sense it is called logic'.[1] Not only did Ramus identify dialectic with logic as a whole (that is, with certain and probable inference alike), but he firmly identified the rules of dialectic with the operations of natural reason. The precepts of dialectic are those which long observation of the processes of natural reasoning as embodied in the discourse of past and present authoritative writers has led men to formulate. Ramus' justification for this claim is that any art, including dialectic itself, is by definition the systematisation of natural operations in some field.[2] The field of operation of dialectic is discourse, and the origin of discourse is natural reason; hence dialectic systematises natural reason.[3] Furthermore, as the systematisation of natural reason, dialectic is concerned with thought rather than with language; grammar systematises language while dialectic systematises the thoughts expressed in language. The success or failure of the traditional handbook is thus for Ramus to be judged by the accordance of its content with the way people do in practice organise their thoughts. And on these grounds he rejects all but the bare Ciceronian framework to which Agricola had originally reduced dialectic for teaching purposes. The parts of dialectic as presented to the student must correspond to the operations of reason; the discussions of terms, predicables, predicaments, analytics and topics in Aristotelian manuals match no such operations; therefore they are irrelevant to dialectic.[4] We thus have with Ramus a treatment of dialectic in the same Ciceronian and Agricolan reformist tradition, but which makes much of a supposed scientific equivalence of dialectic with reasoning.

Like Agricola, Ramus discards the mediaeval form of the dialectic manual altogether, and divides the contents of dialectic

[1] Petrus Ramus (1515–72). For bibliographic and biographic details see Vasoli, op. cit., pp. 333–589 passim; W. J. Ong, *Ramus and Talon Inventory* (Harvard, 1958). Ramus' definition of dialectic is cited from *P. Rami Dialecticae libri duo, scholiis G. Tempelli Cantabrigiensis illustrati* (Cantabrigiae, 1584), p. 1: 'Dialectica est ars bene disserendi, eodemque sensu logica dicta est.' In early versions of the *Dialecticae Institutiones* Ramus uses the phrases 'virtus disserendi' (*Dial.* 1543, 1546), 'facultas disserendi' (*Dial.* 1556) in place of 'ars bene disserendi'. See Risse, op. cit., p. 125.

[2] *Dialecticae Institutiones* (Parisiis, 1543), fo. 19v.

[3] *Animadversionum Aristotelicarum libri XX* (Lutetiae, 1548), pp. 8 and 10–11.

[4] *Animadversionum Aristotelicarum* ed. cit., pp. 34–5.

into invention and judgment, treating invention first. But where Agricola stressed dialectic's relation to *oratio*, and hence the close affiliation of dialectic and rhetoric, Ramus stresses above all that the aim of dialectical theory is to account for the natural operations of reason.[1] For Ramus invention and judgment correspond to the two natural processes by which the mind considers any question, and it is this which gives the rules of dialectic their general application in areas of discourse traditionally separated as 'dialectical' and 'rhetorical'. The places of invention correspond to the natural classification to which the mind assigns experience. Rules for forming propositions and disposing them in syllogisms or larger units of discourse correspond to the mind's natural operations on its store of material in response to any problem or question.[2] The art of dialectic, the *ars disserendi*, thus approximates 'natural' dialectic, or the actual operations of the mind.[3] For Ramus, as for Agricola, the only material relating to discourse as a whole which this study does not cover is that concerned with ornamentation and delivery, and this is assigned to rhetoric.[4]

In the later vernacular editions of the *Dialectica* Ramus illustrates the places of invention and figures of the syllogism with passages of prose and poetry of the period.[5] These illustrations are intended to emphasise the natural and universal character of his dialectical theory. The fact that literary authors, who are not

[1] Ramus' fullest account of dialectic's correspondence with the rules of natural reason is in the first edition of the *Dialecticae Institutiones* (1543).

[2] 'Quod si Dialectica sit ars nativae rationis bene et via quadam instituendae: omnia Dialecticae praecepta ad bene disserendum referentur. Nam quod nativae rationis facultate efficitur, ad id ars Logica tota referetur: id solum intuebitur: id illi finis erit. At actiones illae inveniendi ac judicandi argumenti, quas singillatim persecutus sum, naturae rationis viribus efficiuntur. Itaque universa Logicae disciplinae praecepta ad actiones cogitandi et disponendi argumenti, id est, ad bene disserendum perpetuo et solummodo referentur. Quid enim aliud est bene disserere, quam et invenire causas, effecta, subjecta, adjuncta, aliaque argumenta, et eadem inventa ad judicandum conjungere, id est, axiomate enuntiare, syllogismo disponere, methodo collocare.' Temple, *Dialectica*, ed. cit., pp. 3–4.

[3] Ramus, *Dial.* (1543), fo. 5v-6r.

[4] 'Dialectica mentis et rationis tota est, rhetorica et grammatica sermonis et orationis: Dialectica igitur inventionis, dispositionis, memoriae...artes proprias habebit...Rhetoricae igitur ex sermonis et orationis cultu partes duae solae propriae relinquentur, elocutio et actio.' Ramus, *Scholae rhetoricae* in *Scholae* (Basileae, 1569), cols. 339–40. See also Temple, *Dialectica*, p. 10. Contrast Caesarius' traditional view of rhetoric: 'Cuius [scil. Rhetoricae] partes ab omnibus numero quinque traduntur, Inventio, Dispositio, Elocutio, Memoria et Pronunciatio.' *Rhetorica* (1565 ed.), fo. B7r. Melanchthon, however, adopts Agricola and Ramus' position: 'Rhetorica addit elocutionem inventis a dialectica, et velut ornamentis verborum et sententiarum vestit.' *Opera* (1541), vol. V, p. 172.

[5] See *Dialectique* (*1555*), ed. Dassonville (Geneva, 1964), which reproduces both the 1555 and 1576 texts.

on the face of it concerned with 'artificial' argument, employ forms of reasoning and procedures for selection of material which conform to the rules of the reformed dialectic supports Ramus' claim that these rules do indeed correspond to the 'lois naturelles du raisonnement'. According to Ramus, the task of the dialectician is *descriptive*; to generalise the procedures used by the greatest poets, philosophers and mathematicians.[1] The illustrations have the effect of emphasising this aspect of Ramist dialectic. Natural dialectic is used unconsciously by the greatest minds; the Ramist system will enable more mediocre minds to reach the same heights of expression artificially.

Ramus shares Agricola's concern with the laying out of material in units of discourse larger than the syllogism. For both authors the need for some such discusion arises naturally out of the refocussing of their presentation of dialectic as a whole. But the different emphasis of Ramus' dialectic is reflected in the way in which he handles this topic. In the earliest editions of the *Dialecticae Institutiones* he divides judgment into three types, corresponding to three motives for organising propositions obtained by topical invention. All three, according to Ramus, match natural mental processes.[2] *First judgment* lays out propositions so as to assess the truth or falsity of *quaestiones*, and is identical with *ratiocinatio*, that is to say, with syllogistic.[3] The purpose of *second judgment* is to lay out whole sequences of propositions so as to cover the available material as clearly, concisely and comprehensively as possible. This is achieved by definition of the subject and subsequent presentation by division of all material relevant to it.[4] In this way each proposition will be as closely and naturally as possible related to its predecessor, and nothing (on Aristotle's authority)[5] will be omitted.[6] The ordering

[1] See *Petrus Ramus Dialecticae Partitiones et Institutiones* (1543), fo. 1–3. cit. Dassonville, op. cit,. p. 24. For a sympathetic account of the principles of Ramus' vernacular dialectic, with its descriptive emphasis, see Dassonville, chapter II, 'Les principes de la dialectique ramiste', pp. 20–7.

[2] 'Iudicium sequitur, pars artis maxima, nobilissimaque...Iudicii igitur tres gradus ad eandem naturae matris imitationem, tres, optime, distinguentur.' Ramus, *Dial.* (1543), fo. 19v-20r.

[3] 'Primum itaque iudicium est doctrina unius argumenti firme, constanterque cum quaestione collocandi: unde quaestio ipsa vera, falsave cognoscitur. dispositio autem ipsa, collocatioque syllogismus appellatur: nec quicquam primi iudicii, et syllogismi nomina differunt.' *Dial.* (1543), fo. 20r.

[4] See below, p. 46.

[5] *Posterior Analytics* II, 13.

[6] '[Iudicii gradus] secundus...collocationem tradit, et ordinem multorum, et variorum argumentorum cohaerentium inter se, et perpetua velut catena vinctorum, ad unumque

obtained by application of first and second judgment will, according to Ramus, mirror precisely our perception of natural order, and hence, incidentally, provides the most easily memorizable order of discourse.[1] *Third judgment* transcends both these types of judgment, and aims not simply at mirroring as closely as possible our view of nature, but at grasping the true ends of things as intended by God. This highly metaphysical and Platonist notion[2] is a process of meditation, by means of which the soul takes leave of the constraints of speech and ascends to forms. The account of third judgment, which makes heavy use of Plato's fable of the cave, is by no means clear in its details. But it appears that first and second judgment, supported by the training of the other liberal arts, in particular mathematics and music, provide the mind with the 'simulachra' which are the starting point for third judgment.[3]

In subsequent editions of the *Dialectica*, in the face of severe criticism by more orthodox dialecticians, Ramus substantially moderated his claims for dialectic, and discarded third judgment altogether. Judgment still corresponds to the natural reasoning process, but it now consists of formation of propositions (*axiomatic judgment*, transferred from invention),[4] and two types of *dianoetic judgment*,[5] syllogistic and *method*. Axiomatic judgment forms propositions which are self-evidently true or false from terms assembled and classified under the places of invention (for Ramus, being *false* is equivalent to being ill-formed). Syllogistic combines *axiomata* to determine the truth or falsity of a *quaestio* (a proposition whose truth or falsity is not immediately evident). *Method* is the term adopted by Ramus for second judgment, or ordering of material by division, starting from the definition of the subject.[6] This method is, according to Ramus,

certum finem relatorum: cuius dispositionis partes duae principes sunt, definitio, distributioque: res enim primum universa definienda, et explananda: deinde in partes diducenda est.' *Dial.* (1543), fo. 27r.

[1] *Dial.* (1543), fo. 19v-20r.

[2] The origin for third judgment is clearly Plato's discussion of dialectic in *Republic* VII.

[3] See for instance *Dial.* (1543), fo. 42v; *Anim. Arist.* (Parisiis, 1543), fo. 64r.

[4] In the 1556 edition and later. In the 1546 edition *enuntiatio* is still part of invention.

[5] The term 'dianoetic', which Ramus derives from Plato, is first used in the 1572 edition, but the distinction is made in 1566.

[6] Goveanus had already identified Ramus' second judgment with μέθοδος in Greek authors. He defended Aristotle's omission of its treatment on the grounds that such a study is concerned with order of teaching ('ordo docendi'), whereas Aristotle was interested in the precepts of discourse ('praecepta disserendi'). See *Antonii Goveanii pro Aristotele responsio adversus Petri Rami calumnias* (Parisiis, 1543), fo. 48v.-49r.

the unique one for laying out the available material on any subject whatsoever, under all circumstances. All presentation must begin with what is 'better known', which for Ramus is indisputably what is most general, and proceeds by division to what is less well known (precepts and details).

In his *Animadversiones Aristotelicae* Ramus explicitly rejects Aristotle's remarks that what is nearest to the sense is 'better known to us', and hence may be more suitable as the basis for teaching a novice.[1] He denies that what we first perceive is in any sense 'better known'. However our perceptions of individual instances may contribute to our eventual grasp of the general, the general remains 'better known' or 'prior' intellectually, just as a torch is brighter than and 'prior' to the taper with which it was kindled.[2] Moreover, Ramus insists that the actual procedure by which we come to know universals from singulars is quite irrelevant to dialectic, which is concerned with the presentation of ideas which are already fully formed.[3]

Ramus credits Plato with having originated the use of definition and division as the core of dialectic.[4] There are (as Ramus points out) extended discussions of the technique in the *Sophist* and *Politicus*, and the procedure of many of the dialogues is based upon it. A typical Platonic example from the *Sophist* is the attempt to use this 'methodos' to define 'angling'. The stages of the division are as follows:

An angler is a man with an art;
all arts are acquisitive (getting knowledge or gain) or productive (of welfare or artefacts);
angling is clearly acquisitive;
acquisitive arts are either voluntary exchanges or coercive;
coercive arts fall into fighting or hunting;
angling falls under hunting;
hunting is of the lifeless (diving, etc.) or of the living;
hunting of the living is of walking or swimming creatures (the latter includes birds);
hunting of swimming creatures is fowling or fishing;
fishing is by snaring or striking;

1 See above, p. 34. 2 *Animadv. Arist.* (1548 ed.), pp. 191–2.

3 *Animadv. Arist.* (1548 ed.), p. 202: 'Interponitur decimumquartum caput [*Posterior Analytics* I, 13] de notitiarum in animis per sensus impressione: quod Platonici falsum existimant: quoniam multae sunt notitiae rerum nostris animis non per sensus acceptae. Atque id praeceptum in dialectica proprium locum nullum habet: nihil est inventionis, vel dispositionis.'

4 E.g. *Dial.* (1543) fo. 3v.

striking is by a downward blow (harpoon) or by upward thrust (angling).
Hence angling is an upward striking, fishing, hunting, coercive, acquisitive art.[1]

Plato evidently believed that this technique can be used to establish the essential nature of concepts. In *Prior Analytics* I, 31 Aristotle points out that as a form of inference the divisive method is circular. However, in *Posterior Analytics* II, 13 (96b25f.) he acknowledges that it is a useful supplementary technique in laying out the contents of a subject, because it is the only way of ensuring complete coverage (96b35). In practice, as Ramus and the Ramists use the divisive method, it has only a loose correspondence with the divisive method discussed by Plato and Aristotle, since it is used as a way of organising material under headings, rather than as a rigorous way of formulating definitions. In the presentation of an art, such as dialectic itself, this means beginning with the definition of that art, and an explanation of that definition ('dialectic is the art of discourse'), followed by the definitions of its immediately subordinate genera (invention and judgment), with explanations and examples.[2] In an oration or set theme an even less rigorous version of the divisive technique is used (distinguished in the early versions of the *Dialectica* as the 'prudential' method); the orator starts with a general proposition about his subject which is immediately acceptable to his audience, and continues with successive subdivisions of that.

In its final form Ramus' scheme for dialectic is highly reminiscent of that of Agricola. The striking difference is that in keeping with his original ambition to reconstruct the workings of the mind in the rules of dialectic, Ramus claims complete universality for his system. Where Agricola aimed at providing a general guide to elegant discourse, Ramus insists on the appropriateness of his places, his rules for syllogisms, and above all his divisive method, in activities as diverse as the laying out of the material of astronomy and the composition of a Virgilian epic. The difference between these areas of application of the 'ars disserendi' lies not in the techniques to be used (since all discourse is governed by the same rules – the rules of natural

[1] Paraphrased from *Sophist* 218E-221C.
[2] For a clear account of this loose procedure see *Dial.* (1543), fo. 28v.

reason), but only in the status of the propositions which serve as the basis for discourse. It is a central tenet of Ramist dialectic that demonstrative or certain inference such as is deployed in teaching the arts is distinguished from plausible or probable argument such as is used in literature and oratory merely by the fact that the premises of the former conform to the three Aristotelian criteria for the scientific certainty of principles.[1] Much of the controversy which arose amongst dialecticians on the subject of dialectical method was the result of the inflexibility and exclusiveness which are of the very nature of the Ramist approach. The topic of dialectical method arose naturally as a result of the shifted emphasis in dialectic teaching within a particular tradition, and was always confined to the sphere of teaching what is already clearly grasped by the teacher himself, and beyond question. The eccentricity of Ramus' approach, which survived his later concessions to more orthodox dialecticians, was to claim that his vastly simplified and schematic account of dialectic is ideally suited to all discourse because of its conformity with the mental process by which we assimilate knowledge.[2] Equally eccentric was his claim that the Platonic method of definition and division (Galen's *diaeresis*) is the unique way of presenting any material, in any field, because it matches our understanding of the true divisions of nature.

Methods of demonstration

The dialecticians discussed in the previous section were predominantly concerned with the orderly presentation and display of material by orators and teachers. Because of their concern with teaching, and their view that for the purposes of dialectic the principles of a subject are taken on trust from the specialists in the field, or are based on consensus of opinion, they were not

[1] These criteria, based on Aristotle's three rules for scientific propositions in *Posterior Analytics* I, 4, became known as the rules of 'truth', 'justice' and 'wisdom'. See below, p. 52. See Ramus' correspondence with Scheckius, in *Petri Rami...et Audomari Talaei collectaneae prefationes, epistolae, orationes...*(Marpurgi, 1599 ed.), pp. 176 and 181 (for example), where Ramus' main case against Scheckius' separation of demonstration and dialectic is that it is unnecessary to distinguish separate fields for the certain and the plausible, since all that needs to be added for the former is the use of the three rules for the initial propositions.

[2] This emphasis on the conformity between the rules of Ramist dialectic and mental operations, rather than simply (as in grammar) between these rules and the observed relations between terms and propositions, recurs repeatedly in Ramist handbooks. See for instance, Temple, *Dialectica*, p. 66.

bothered by any fundamental questions about the ultimate foundations of knowledge. *Demonstration*, or reasoning leading to scientifically certain conclusions ('syllogismus faciens scire', 'a syllogism producing knowledge'), is treated by the dialecticians, where they handle it at all, simply as a stronger counterpart to syllogistic based on plausible premises.[1]

Peter of Spain's *Summulae* had omitted treatment of the *Posterior Analytics* and demonstration altogether. Some of his commentators claimed that this was because the topic was considered too difficult for students (and the text was of comparatively new discovery when the textbook was written); others maintained that it lay outside the scope of a dialectic or disputation. Thus even before Agricola's determined efforts to build a dialectic of teaching and eloquence there was hesitation about the place of scientific inference in dialectic teaching.

Both Caesarius and Melanchthon incorporate a discussion of demonstration into their dialectic manuals, to complete the corpus of Aristotle's Organon (Seton, however, does not). And both these writers ignore the fact that the *Posterior Analytics* is partly about the acquisition of *principia* (which they place outside the scope of dialectic), and extract from Aristotle's work only such material as may readily be applied to discussion of the *consequences* of principles already established.[2] Demonstration is 'a syllogism in which, either from principles known by nature, or from universal experience, or from prior definition, we extract consequences consistent with these, and a necessary and immutable conclusion; or we show appropriate effects to follow from proximate causes, or vice versa'.[3] Both authors stress that demonstrative reasoning must be based on *principia*,

[1] Valla made this something of a point of principle, on the grounds that it was the emphasis on demonstration which led logicians to separate the study of dialectic from the other trivium subjects. See L. Valla, *Dialecticae Disputationes* (1439) in *Opera* (Basileae, 1540), I. As we saw, Ramus insisted on a single dialectic for certain and probable inference, on similar grounds.

[2] This is the characteristic dialectician's way of dealing with the difficulty. See, for instance, Trapezuntius' Latin synopsis of the *Posterior Analytics*, in which he picks out precisely the passages which Caesarius and Melanchthon draw upon. See also J. Eckius, *Elementarius Dialectice* (Augustae Vindelicorum, 1517). fo. C6v-D1v.

[3] Melanchthon, *Erotemata Dialectices* (Wittebergae, 1555), p. 243: 'Demonstratio est syllogismus, in quo aut ex principiis natura notis, aut universali experientia, aut ex definitione de subsumpto, bona consequentia, necessariam et immotam conclusionem extruimus, aut ex causis proximis effectus proprios sequi ostendimus, aut econtra procedimus.' See also Raiianus in Caesarius, *Dialectica* (Coloniae Agrippinae, 1568), fo. S1r: 'Demonstratio est syllogismus constans propositionibus primis, veris, immediatis, notioribus, et causam conclusionis continentibus.'

or certain, indemonstrable truths, and characterise types of principle. But both emphasise that principles are self-evidently true; their truth is apparent by common consensus and immediate scrutiny. The ultimate source of principles is not in question.[1]

Caesarius and Melanchthon associate their discussion of demonstration procedures with *a priori* and *a posteriori* methods of proof used by geometers. This helps to obscure the need for any discussion of an independent procedure for the discovery of principles. In geometry there appears to be no difficulty. The axioms or principles are either definitions ('A signe or point is that, which hath no part'; 'A line is length without breadth'; 'A superficies [surface] is that, which hath onely length and breadth'[2]), or incontrovertible ('Thinges equall to one and the self same thyng are equall also the one to the other'; 'Every whole is greater then his part'[3]). Both authors characterise the *methods* which are used in demonstration as procedures which resemble the proof techniques of geometry, and establish a chain of inferences of guaranteed certainty by commencing with, or concluding with, 'immediate principles, which are known in themselves, and do not require to be established by other means'.[4]

For Caesarius, the two types of demonstrative method are two of the forms of demonstrative syllogism discussed by Aristotle in *Posterior Analytics* I, 13.[5] Both these syllogisms, according to Caesarius, infer an effect from its cause. The one (demonstration *propter quid*) infers an effect from a cause which is 'primary' and 'immediate'; that is, which holds essentially when and only when the effect occurs. The other (demonstration *quia est*) infers an effect from a cause which is not linked to it by definition, but which nevertheless provides valid grounds for the conclusion. Caesarius' example is that in the syllogism:

> what breathes has lungs
> trees have no lungs
> therefore trees do not breathe

[1] See Melanchthon, *De Dialectica* (1531 ed.), fo. L4v-L5v. See also *Erotemata Dialectices* (1562 ed.), pp. 233–7; Caesarius, *Dialectica* (1568 ed.), fo. S1r-S4r; fo. T2v-T4v.

[2] The examples are from H. Billingsley, *The Elements of Geometrie of the most auncient Philosopher Euclide of Megara* (London, 1570), fo. 1r and fo. 2r.

[3] Billingsley, *Euclid*, fo. 6v and fo. 7v.

[4] Melanchthon, *Erotemata Dialectices*, ed. cit., p. 244.

[5] See below, p. 55.

'not having lungs' is the 'primary' and 'immediate' cause of 'not breathing', and hence provides a demonstration 'propter quid' of the effect. On the other hand, in the syllogism:

what is an animal has lungs
trees have no lungs
therefore trees are not animals

'not having lungs' is not the immediate cause of 'not being an animal' and hence the demonstration is 'quia est' only. This weaker form of demonstration, according to Caesarius, is characteristic of the 'subaltern' or applied arts, which use mathematical principles (say) to prove an effect in a practical field like perspective.[1] In this account Caesarius follows the tradition of commentators on the *Posterior Analytics* in identifying methods of demonstration strictly with forms of syllogism.

Melanchthon, on the other hand, associates methods of demonstration with more extended inference sequences, and draws together the geometers' discussion of such methods and Galen's remarks about method to which I alluded earlier.[2] The passage in which he discusses these is an important one:

> The best known names [for these methods], which were used by the geometers are 'composition' [or] 'synthesis', which proceeds *a priori*, and on the other hand, 'resolution' or 'analysis' which returns *a posteriori* to principles. Nor is there any doubt that these names signify the same thing in Galen as in the geometers, when he says that there are three ways of teaching, resolution, composition, and definitions. Resolution is, for instance, when from the symptoms of a disease we look for its seat, affect and causes. On the other hand, composition is when we first describe the parts of the body, then the causes of the disease, then its symptoms. But what we call 'definitions' are rules and definitions which have not been demonstrated, like aphorisms.[3]

[1] Caesarius, *Dialectica* (1568), fo. S3v-S4r. Caesarius' commentator Raiianus points out that demonstration 'quia est' is more conventionally interpreted as inferring a cause from its effect.

[2] See above, p. 40.

[3] *Erotemata Dialectices*, ed. cit., p. 243: 'Geometris usitata nomina sunt, et notissima, compositio Synthesis, quae a priore procedit, Econtra resolutio seu Analysis, quae a posteriori ad principia regreditur. Nec dubium est, hae appellationes apud Galenum idem significare, quod apud Geometras, cum ait tres esse doctrinarum vias, resolutionem, compositionem, et definitiones. Resolutio est, ut cum ex signis morbum, locum, affectum, et causas quaerimus. Econtra compositio, cum initio corporis partes describuntur, deinde morborum causae, postea signa. Sed definitiones vocantur regulae et definitiones sine demonstrationibus, ut aphorismi.'

What is striking about this passage is the reckless oversimplification involved in lumping together methods of geometrical proof, methods of demonstration in science and Galen's instructions for teaching the practical art of medicine.

Melanchthon's account is less important as a direct influence than as an indication of the sort of muddles dialecticians could get into on the subject of method.[1] As we saw, the position of Galen's three methods was always ambiguous. Ostensibly teaching methods, they tended to be confused with methods of discovery, and (as in the present case) it is often not at all clear whether in the context of demonstration what is being discussed is presentation of a subject starting with or ending with self-evident principles, or independent derivation of principles and deduction of further consequences from principles. Averroës' widely cited introduction to his commentary on Aristotle's *Physics*, in which he discusses the principles according to which Aristotle laid out that work, is similarly ambiguous.[2] Such accounts led a whole series of renaissance writers to discuss method in a way which does not belong clearly either to the tradition of teaching method in dialectic, or to the tradition of discussion of demonstration in the sciences (although the two contexts had been quite satisfactorily separated by Grosseteste much earlier). It was only towards the end of the sixteenth century that attempts were made by such writers as Zabarella, Pacius and Scheckius[3] to separate the various contexts of discussion of *ordo* (method of presentation) and *methodus* (method of investigation).[4]

At one point only in his discussion of demonstration does Caesarius allude to any procedure prior to the demonstration of consequences of self-evident *principia*. He suggests that in the acquisition of certain knowledge, demonstration is preceded by 'definition' and 'division'.[5] Glareanus comments on this remark:

[1] Although, in the present context it should be remembered that Melanchthon was specified as a basic teaching text by the Cambridge statutes of 1535. It is striking that Melanchthon openly commits all the errors which Bacon complains about of dialecticians, and which Seton (for instance) only implicitly makes.

[2] *Aristotelis de physico auditu libri octo, cum Averrois Cordubensis variis in eosdem commentariis*...(Venetiis, 1562), fo. 4r.

[3] Jacobus Scheckius (1511–1587); Julius Pacius (1550–1635); Jacopo Zabarella (1533–1589). On Pacius see Gilbert, op. cit., p. 192.

[4] Zabarella deals explicitly with the confusion by Averroists of *ordo* and *methodus*. *De Methodis*, in *Jacobi Zabarellae Patavini opera logica* (Francofurti, 1608), col. 133D. See also Scheckius, *De Demonstratione libri XV* (1564).

[5] *Dialectica*, ed. cit., fo. R7r-R7v.

> The first operation of the intellect corresponds to *definition* which is simple apprehension. The second operation corresponds to *division*, which is composition and division. The third operation of the intellect corresponds to *demonstration*, or the discourse of reason.[1]

This observation is based on Aristotle's remarks on definition and division in the second book of the *Posterior Analytics*, to which some Aristotelian commentators attached considerable weight. However, in his extended discussion of definition and division in the eighth book of his *Dialectica*, Caesarius gives no indication of how these techniques are envisaged as establishing principles, and confines himself to a traditional account of definition and division in the context of disputation, based on Aristotle's *Topics*.

Ramus takes a stand which whilst obviously misguided is in line with the tradition of treatment of demonstration in dialectic handbooks. Caesarius, in his discussion of demonstration, invoked Aristotle's three criteria for the scientific status of propositions (that they be true primarily, essentially and universally[2]) as providing a way of identifying a syllogism as demonstrative.[3] Ramus apparently believed that given definitions fulfilling these criteria, and combining them with presentation according to his divisive method, it follows of necessity that one will obtain a complete, unique coverage of all and only such material as is relevant to the subject (in particular, the academic discipline) under discussion. For Ramus, the rule κατὰ παντός (true universally, 'de omni', or the rule of truth) ensures that all the precepts displayed are necessarily true, i.e., in all instances; the rule καθ' αὑτό (true essentially, 'per se', or the rule of justice) ensures that all and only such material as strictly belongs to a specified field is included (grammar does not include precepts from rhetoric, for instance); and the rule καθόλου πρῶτον (true primarily, 'universaliter primum', or the rule of wisdom) ensures that the precepts handled are compatible (general precepts are not intermingled with particular cases, and so on).[4]

[1] *Dialectica*, ed. cit., fo. R7v-R8r.

[2] See below.

[3] *Dialectica*, ed. cit.

[4] The clearest accounts Ramus gives of the three rules and their use occur in the prefaces to his textbooks on grammar, dialectic, physics, etc. See for instance, the preface to the first edition of the physics textbook: 'Logicae leges illae de materia formaque artis ante oculos habendae sunt, ut materia omnis sit κατὰ παντὸς, καθ' αὑτο, καθ'ὅλου πρῶτον de omni, per se, universaliter primum. Prima lex est veritatis, ne ullum sit in arte documentum, nisi

Effectively, Ramus takes the third Galenic method as described by Melanchthon (presentation by 'rules and definitions which have not been demonstrated'), and adds criteria for selecting the material to be thus presented, to yield (so he claims) a 'scientific' procedure which guarantees the comprehensiveness and validity of the content.

In his *De Demonstratione libri XV* (1564), and in a published correspondence with Ramus, the Aristotelian commentator Scheckius attacked Ramus' lack of awareness of the need for a separate discussion of the procedures by which initially we come to know. Challenging Ramus' claim that logical demonstration and dialectical reasoning are all one, since all material, certain or probable, can be handled by the rules of dialectic, Scheckius points out that the difference between the two fields lies in their function, not in the type of material they handle. Dialectic is used for organising material for teaching and disputation; demonstration is concerned with the discovery of fundamental principles and their consequences. It is therefore demonstration which corresponds to invention, and dialectic to judgment or disposition. This is a far more telling distinction, according to Scheckius, than that made by Ramus between dialectical invention according to the places, and judgment by syllogism and method. The three rules do not endow the divisive method with demonstrative validity; under all circumstances this method is merely an aid to teaching. In demonstration, according to Scheckius, the method of Aristotle's *Posterior Analytics* is the method for establishing necessary truths.[1]

The work in which Scheckius originally expounded his views on procedures for construction of the arts and sciences, and pro-

omnino necessarioque verum. Itaque non modo falsa, sed fortuita tollentur. Secunda lege cavetur amplius, ut artis decretum sit non tantum omnino, necessarioque verum, sed homogeneum, et tanquam corporis ejusdem membrum...Haec justitiae lex est ad regendos artium fines, et suum cuique tribuendum, justissima. Tertia demum lege sancitum est, ut artis praecepta non sint duntaxat omnino necessarioque vera, nec homogenea tantum, sed propria et partibus reciproca: neque generale speciei, aut speciale generi tribuatur, sed generale generaliter, speciale specialiter exponatur...Haec tertia lex est sapientiae. De forma lex unica est, ut absolute notius et clarius antecedat...Hae sunt Aristotelis de materia formaque artium logicae leges, communes omnium temporum, locorum, hominum, nullis exceptionibus consilii, voluntatis, occasionis cujusquam astrictae.' Reprinted in Petri Rami ...*Audomari Talaei collectaneae prefationes, epistolae, orationes...* (Marpurgi, 1599 ed.), pp. 49–50. See also the prefaces to the fourth edition of the grammar textbook, and to the second edition of the dialectic textbook, reprinted in the same work, pp. 8–11 and pp. 31–4.

[1] See, for instance, Scheckius' reply to Ramus, printed in *Petri Rami...Audomari Talaei collectaneae prefationes, epistolae, orationes...*ed. cit., pp. 185–93.

cedures for establishing apodeictic truths (before he was drawn into controversy with Ramus), the *De Demonstratione*, is essentially a commentary on the *Posterior Analytics*. It belongs to the tradition of commentaries which starts with the commentators of late antiquity (especially Themistius), and is continued through the influential commentaries of Averroës, Grosseteste and Walter Burleigh. In these commentaries attempts are made to systematise the *Posterior Analytics* to make clear Aristotle's supposed recipe for discovery of necessary truths.[1]

The three main themes of the *Posterior Analytics* which provide fuel for the renaissance commentators' discussions of methods of discovery are *demonstration, definition* and *induction*. According to modern readings of the *Posterior Analytics*,[2] Aristotle sees induction from particulars perceived by the senses as the source of all fundamental principles (*Posterior Analytics* II, 19). Syllogistic demonstration derives the consequences of principles, and methods of manipulation of definitions are dismissed as question-begging (for example, *Posterior Analytics* II, 5; II, 13; *Prior Analytics* I, 31). Many renaissance commentators read the work differently. They tended to minimise the rôle of the induction from sense perceptions described in *Posterior Analytics* II, 19, as they regarded induction as a weak and fairly unreliable form of inference, and they tended to substitute as accounts for the discovery of fundamental principles a procedure built out of the syllogistic demonstrations of causes from effects and effects from causes described in *Posterior Analytics* I, 13 and II, 8.

For example, in his *De Methodis*, Jacopo Zabarella makes a

[1] Gilbert (op. cit., pp. 158–63) recognises the importance of such works as Scheckius' as the 'scientific' counterparts of the 'arts' discussion of method, but he incorrectly speculates that sixteenth-century Aristotelians 'were goaded by the innovations of Ramus and others to return to the Greek text of Aristotle and to seek there the answer to the challenge of the Humanists'. It is clear (for example, from A. C. Crombie's analysis of Grosseteste's commentary, *Robert Grosseteste and the origins of experimental science 1100–1700* (Oxford, 1953)) that the western discussion of method of discovery in the sciences antedates the humanist discussion of teaching method by at least four hundred years. Indeed, from Ramus' remarks on the inadequacy of existing methods when applied in dialectic it is perhaps reasonable to infer that the dialecticians consciously developed their methods of presentation as a counterpart for the thoroughly established tradition of methods of discovery in the sciences, and that Sturm and Melanchthon adopted the term 'method' for such discussions by analogy with its use in discussions of demonstration in the sciences.

[2] G. E. R. Lloyd, op. cit., pp. 122–7; R. McKeon, 'Aristotle's conception of the development and the nature of scientific method', *Journal of the History of Ideas*, 8 (1947), pp. 3–44; H. D. P. Lee, art. cit.

clear distinction within 'methodus sensu lato' (method in the broad sense, which covers all mental processes for acquiring knowledge, and for presenting existing knowledge) between teaching methods ('ordines') and methods of discovery in the sciences ('methodus sensu stricto', method in the narrow sense).[1] His discussion of 'ordines' runs through the main alternative methods put forward by the dialecticians, including the ones we have considered above. His treatment of 'methodi' (discovery techniques) is based on close, meticulous reading of the *Posterior Analytics*.[2] He maintains on the strength of this reading that there are two methods of discovery only, compositive syllogism and resolutive syllogism, otherwise called 'demonstratio propter quid' and 'demonstratio quod est' (more usually called 'demonstratio quia').[3]

Demonstration *propter quid* infers an effect from its immediate cause by a syllogism. We saw Caesarius' example above; the example which Aristotle gives in *Posterior Analytics* I, 13 (78a22-79a16) is:

> what is near does not twinkle
> the planets are near
> therefore the planets do not twinkle

Here 'nearness' is the cause of 'not twinkling'; hence the syllogism infers an effect from its cause. 'Nearness' is the immediate cause because (so Aristotle claims) nearness is not just true of heavenly bodies which do not twinkle, but is peculiar to them. Because 'nearness' is peculiar to 'bodies which do not twinkle' it

[1] See *Jacobi Zabarellae Patavini opera logica* (1608), cols. 138–9. I have used Zabarella as my example (both his *De Methodis* and his commentary on the *Posterior Analytics*), for a variety of reasons. In the first place, he gives a particularly clear account of methods of discovery, and their distinctness from methods of presentation, and gives unusually full exemplification. In the second place, authors, like Crombie, who have suggested that there are links between Bacon's inductive method and renaissance methods of demonstration (a view which, as I have said, I do not follow) have singled out Zabarella as marking the high water point of renaissance Aristotelianism. It therefore seemed desirable to allow the reader to compare Zabarella's approach with Bacon's directly. Finally, both the commentaries on the *Posterior Analytics* which circulated in England, and were finally published there, in the sixteenth century, state their indebtedness to Zabarella (see Kosman, op. cit., p. 164). On the other hand, Zabarella is more of an Aristotelian purist than is entirely typical of the period. He is apparently anxious to cleanse Aristotelian logic of the taint of the practical arts, and hence to withdraw from positions adopted by more liberal commentators like Baldwin.

[2] Zabarella correctly identifies the three Galenic methods as strictly teaching methods, i.e. *ordines* (ed. cit., cols. 165 and 166).

[3] For the claim that 'methodus sensu stricto' is syllogistic see cols. 226–9. See also Zabarella's commentary on the *Posterior Analytics*, I, XII.

is possible to reverse the major premise to get another syllogism as follows:

what does not twinkle is near
the planets do not twinkle
therefore the planets are near

This is demonstration *quod est* or *quia*, demonstration of the fact rather than of the reasoned fact (as we saw, Caesarius interprets demonstration *quia* quite differently). It demonstrates a cause (the nearness of the planets) from its effect (the planets do not twinkle as a result of their nearness).

Aristotle made it clear in his discussion of these two forms of demonstration that the major premise is got by induction. Demonstration *quod est* (demonstration of the fact) plays little rôle in the strategy of the *Posterior Analytics* as a whole, but for many renaissance commentators[1] it is the mainstay of discovery in the sciences.

Zabarella, in common with other Italian commentators, tried to get round the obviously question-begging nature of demonstration *quod est* as a tool of discovery by claiming that the two forms of demonstration may legitimately be combined so as to yield a method (the *regressus*) which is not viciously circular. His argument in the *De Regressu* is that when we first demonstrate a cause from an effect we cannot be sure that the cause postulated is the immediate cause of the effect. However, by a process of cogitation we come to realise that the major premise is indeed the proximate or immediate cause, and hence that the effect can be demonstrated with certainty from the cause by demonstration *propter quid*.[2]

[1] For instance, for many of the commentators cited by Risse and Schüling (op. cit.).

[2] Zabarella's account of the *regressus* is as follows: first the cause is derived from the effect (*De Regressu*, chapter IV, *Opera logica*, ed. cit., cols. 484–6); then follows *consideratio*, by which the cause is judged to be unique, and finally demonstration *propter quid* (of the effect from the unique cause) (*De Regressu*, ed. cit., cols. 486–9). These three stages may not be a temporal succession, according to Zabarella, but only a logical one. Kosman has given a careful account of Zabarella's treatment of *regressus*, op. cit., pp. 116–38.

J. H. Randall has claimed that the Aristotelian discussion of demonstration and *regressus* anticipates the scientific method of Galileo. See Randall, 'The development of scientific method in the school of Padua', *Journal of the History of Ideas*, I (1940), pp. 177–206, reprinted in *The School of Padua and the emergence of Modern Science* (Padua, 1961), pp. 15–68. Crombie (op. cit., chapter XI) repeats many of Randall's views. Randall and Crombie fail adequately to discriminate the commentators' treatment of syllogistic demonstration from their treatment of methods for handling definitions. Because both involve an ascent and a descent they become elided. Both authors regard the distinction between ascending methods (demonstration *quia* or *quod est* and composition of defini-

Zabarella notes that some commentators have proposed methods of definition (based on the second book of the *Posterior Analytics*, which they mistakenly suppose to be entirely about definition) for arriving at fundamental principles.[1] These methods of definition are the methods of division and composition discussed by Aristotle in *Posterior Analytics* II, 13, which Caesarius, as we saw, favoured as the basis for knowledge. I have already described logical division in the context of dialectical teaching method.[2] The converse method of composition is the ascent from the most evident classes of objects, *infimae species*, to definitions of more inclusive classes. Given the definitions of known subclasses of an inclusive class, the investigator eliminates all but the common components of these definitions to arrive at a single general definition of the class to which all the subclasses belong. Aristotle's example from *Posterior Analytics* II, 13 (97b15–22) is the following:

> If we are inquiring what the essential nature of pride is, we should examine instances of proud men we know of to see what, as such, they have in common; e.g. if Alcibiades was proud, or Achilles and Ajax were proud, we should find, on inquiring what they all had in common, that it was intolerance of insult; it was this which drove Alcibiades to war, Achilles to wrath, and Ajax to suicide. We should next examine other cases, Lysander, for example, or Socrates, and then if these have in common indifference alike to good and ill fortune, I take these two results and inquire what common element have equanimity amid the vicissitudes of life and impatience of dishonour.[3]

Zabarella sticks to Aristotle's *Posterior Analytics* text in rejecting division as essentially question-begging;[4] composition of defini-

tions) and descending methods (demonstration *propter quid* and logical division) as anticipating the orthodox modern distinction between inductive and deductive components in scientific method. My own reading suggests that renaissance discussions of scientific method are too complicated to admit such a tidy parallel. In particular it should be noted that it is unlikely that anything which can be identified with modern views on the inductive component of science is to be found in the writings of such an Aristotelian purist as Zabarella, who believed that the goal of scientific method is incorrigible necessary truths. C. B. Schmitt, *A critical survey and bibliography of studies on renaissance Aristotelianism 1958–69* (Padua, 1971), pp. 38–46, questions Randall's claims.

It is, however, worth noting that one of the commentators on whom Randall rests his case (Augustinus Niphus, 1473–1546) takes what appears to be a genuinely original and independent stand, and is sceptical of the certainty attainable by demonstrative methods. See Risse, op. cit., pp. 218ff.; Kosman, pp. 116–38.

[1] *De Methodis*, ed. cit., cols. 275–6.

[2] See above, p. 45.

[3] *The Works of Aristotle translated into English* (Oxford, 1928), ed. Ross, I.

[4] Ed. cit., col. 1191B, C.

tions he, like many other renaissance commentators, equates with conceptual induction.[1]

The induction of *Posterior Analytics* II, 19 is considered by Zabarella to be a secondary and imperfect form of resolutive method. Whereas demonstration *quod est* can make known to us causes which are in no way apparent from the perceived effects, induction can only make known to us causes which are clearly implicit in the perceived effects.[2] Not all commentators shared his determination to minimise the rôle of induction in discovery, and as we saw, the dialecticians (where they considered the question at all) adopted the view that (in some unspecified sense) the principles of arts and sciences are arrived at by induction. Of particular interest in the context of the present discussion are Seton's remarks on induction and the derivation of principles in his *Dialectica* (which, it should be remembered, was in all probability the manual which introduced Bacon to dialectic at Cambridge).

Although Seton only discusses induction by incomplete enumeration as a supporting device in dialectic, and rhetorical induction as a persuasive device in oratory, he does refer to Aristotle's conceptual induction. Both he and his commentator Carter maintain that induction establishes the truth of *principia* by supporting example, although their truth cannot be demonstrated.[3] And Carter adds that *principia* are derived in the following four ways: by use, by experience, by induction and by intellect. Here Seton and Carter show that, as far as they are concerned, the induction alluded to by dialecticians in passing as participating in the establishing of principles and the induction by incomplete enumeration which provides supporting argument in dialectic, and persuasive force in rhetoric, are one and the same. And their remarks provide us with a perfect example of the approach by dialecticians towards principles against which Bacon launched such blistering attacks. They at once pass off the derivation of principles as by and large outside their concern, and, where they do mention any such procedure, link conceptual induction with a 'puerile' induction by incomplete enumeration.

[1] Ed. cit., col. 264D; col. 1196D.
[2] Ed. cit., col. 269D, E.
[3] *Dialectica*, ed. cit., fo. Nviiiv-Oiiir.

2

An English dialectical controversy

The dispute between the Cambridge dons Everard Digby (*c.* 1550–92)[1] and William Temple (1555–1627)[2] over the nature and uses of dialectical method provides several important clues for the understanding of Bacon's polemical response to dialectic. It is obviously of some general interest as providing a glimpse of some of the key issues which were topics of debate at the time when Bacon attended Cambridge: something which no amount of scrutiny of the standard elementary textbooks can provide. But it has also a more specific relevance. For in certain respects Temple's polemic against Digby runs parallel to Bacon's polemic against the dialectical tradition as a whole.

In the last chapter I emphasised two related features of the reformed dialectic handbook – the tendency to treat lightly or dismiss the problems of acquisition of knowledge, and the tendency, particularly evident in the writings of Melanchthon and his followers, to discuss methods of definition and demonstration in ways which blur the distinction between acquisition and presentation of knowledge. These are precisely the defects which, as we shall see, Bacon finds in the writings of the 'Dialectici'. In the writings of Digby, a professed Aristotelian with mystical Platonist sympathies, these tendencies are carried to an extreme. The acquisition of knowledge by induction is completely ignored, and all possible contexts for the discussion of method are scrambled together in glorious confusion. Temple, in his clear-headed and precise retaliations repeatedly and doggedly insists on two simple points. The contexts of discovery

[1] For biographical details see J. Freudenthal, 'Beiträge zur Geschichte der englischen Philosophie [I]', *Archiv für Geschichte der Philosophie*, 4 (1891), pp. 450–77.

[2] For biographical details see J. Freudenthal, art. cit. [III], *Archiv für Geschichte der Philosophie*, 5 (1892), pp. 1–41.

of knowledge and teaching of knowledge should be carefully separated, and in the context of discovery induction is the sole source of knowledge.

In 1580 Digby published a pamphlet designed to stem the growing tide of Ramism in Cambridge, entitled *De duplici methodo libri duo, unicam P. Rami methodum refutantes*.[1] On the basis of the traditional Aristotelian distinction between 'better known to us' and 'better known in nature'[2] he maintained against the Ramists that in all contexts in which human knowledge and its acquisition are discussed there must be two opposed ordering procedures for material. In teaching, for instance, the novice can comprehend at first only what is 'better known to him' – the immediate perceptions of his senses and products of his own limited experience. Hence the teacher must proceed by composition from the lower species which his student clearly grasps, to the most general precepts of the art which is being considered. These general precepts are the ultimate goal of knowledge.[3] A source for Digby's method of composition for teaching the novice is once again Aristotle's *Posterior Analytics* II, 13. Here as in the introduction to the *Physics* Aristotle recommends composition from immediately apprehended *infimae species* as an easier alternative to approaches which start from definitions of the most general notions. And likewise in *Topics* VI, 4 (141b15-142a5) he concedes that it may be necessary in teaching to start from other than the most general notions. Digby maintains that once the novice becomes expert, and grasps clearly what is 'better known in nature', the general precepts themselves,

[1] The chronology of the debate has become confused in the literature. The correct order of events is as follows: early in 1580 Digby published an open attack on Ramism, *De duplici methodo* (entered in the Stationer's Register 3 May, 1580). A few months later Temple replied under a pseudonym with *Admonitio de unica P. Rami methodo reiectis caeteris retinenda* (Londini, 1580), in which, in the course of his attack on Digby's position, he strongly criticised an earlier work by Digby (not explicitly directed against Ramus), *Theoria Analytica, viam ad Monarchiam Scientiarum demonstrans...*(Londini, 1579). Digby replied late in 1580 with a pamphlet entitled *Everardi Digbei Cantabrigiensis admonitioni Francisci Mildapetti responsio* (entered 3 November, 1580 in the Stationer's Register). Temple published his final contribution to the debate under his own name in 1581, *Pro Mildapetti de unica methodo defensione contra Diplodophilum* (entered 18 May, 1581 in the Stationer's Register). Gilbert and Vasoli both erroneously state that there were only three pamphlets, and that Temple began the exchange. See Gilbert, op. cit., pp. 202–8; Vasoli, op. cit., p. 592.

Everard Digby was not the same person as the Catholic contemporary of the same name who was involved in the gunpowder plot and executed for treason. This error (whose source is Aubrey's *Brief Lives*) is also common in the literature!

[2] See above, p. 34.

[3] *De duplici methodo*, fo. B8v-C1r; fo. C1v-C3r; fo. D2v-D4r.

further material may be presented to him by resolution of the whole subject into lesser parts.[1]

To reinforce his point that there are two opposed starting points for all ordering procedures, Digby considers in turn a variety of methodological pairs, all of which, he claims, derive from his initial distinction. Dialectical methods are two in number: invention and judgment.[2] The handling of *quaestiones* involves two methods: one for a simple theme, another for a composite theme (and either composition or resolution will be the appropriate method for handling a composite theme, depending on whether the auditor is a novice or an expert).[3] Above all, the process of coming to know by experience depends critically upon the distinction between what is clear to us, and what is clear in itself or in nature.[4]

In his earlier *Theoria Analytica* (1579) Digby had considered all handling of definitions and precepts derived from experience (what is prior or better known to us) as merely a preliminary to a mystical Platonist method of *analysis*, by which we ascend from apprehension of forms (the 'mundus medius') to what is absolutely prior, the form of the Good (the 'mundus suprasupremus'). In the *De duplici methodo* he sustains his distinction between the naturally or intellectually prior towards which we strive, and the immediate perceptions 'prior' to us, which form the only basis for our knowledge. But in the pamphlet the mystical process of ascent from forms is alluded to only briefly, on the grounds that it is the final stage in learning.[5] Instead, by sliding from 'scire', to know by learning from a teacher, to 'scire', to come to know from nature, Digby falls back on the methods of composition and division which he proposed for teaching, as methods for acquiring knowledge.[6]

[1] *De duplici methodo*, fo. D4r.

[2] *De duplici methodo*, fo. B7v.

[3] *De duplici methodo*, fo. D7v-E1r. Compare the methods of Willichius and Hemmingius above, p. 39.

[4] 'Sagax natura...hominem ab ortu sic instituit, ut duplicis methodi duplex in se principium representet, lucis scilicet et tenebrarum. Illo universaliter primo, et confuse, divinamus omnes: hoc particulariter et distincte discernimus.' *De duplici methodo*, fo. C3r.

[5] 'Analysis est divina virgula, quae gressus dirigit mortales, ad lucidum veritatis fontem. Haec Homeri aurea est catena, cuius gradibus ascendimus ad Chrystalinum scientiae coelum...Haec methodorum est exactissima, sed nobis obscurrissima et remotissima. Itaque posterius est edocenda.' *De duplici methodo*, fo. C5r.

[6] In eliding as *doctrinal* methods methods of discovery which start with what is better known to us (individuals and *infimae species*) and teaching methods which exploit the fact that what is less general is closer to the student's experience, Digby shows his heavy depen-

In his eagerness to support his bipartite distinction between methods of ascent from particulars and methods of descent from precepts Digby seriously entangles the various discussions of method. In his anonymous *Admonitio de unica P. Rami methodo reiectis caeteris retinenda* (1580), and in a second retaliatory pamphlet published under his own name, *Pro Mildapetti de unica methodo defensione contra Diplodophilum* (1581),[1] William Temple exposed the confusions of his Aristotelian adversary. In the course of doing so he showed what an intelligent commentator could make of the various classical and contemporary discussions of method. Temple draws heavily on Ramus' basic arguments against the Aristotelian position, but he presents these with a clarity and conciseness which are in marked contrast to Ramus' longwinded and apologetic works. He also dispenses entirely with the metaphysical platonising to which Ramus frequently had recourse in his earlier works, and indeed upbraids Digby for using the same sorts of argument in his *Theoria*.[2]

Attacking the core of Digby's argument, Temple asserts that the Aristotelian distinction between 'better known to us' and 'better known in nature' is a useless and misleading one. The only sense in which one item of knowledge can fruitfully be said to be 'prior' to another is if the first is needed to explain the second. This intellectual priority, Temple claims, belongs uniquely to the most general precepts. The more general the precept, the more can be explained in terms of it. Particular sense perceptions are in no sense prior, simply because our apprehension of them precedes in time our apprehension of the general notions which they instantiate, any more than the seed is

dence on the commentary of Louvain, which he frequently cites (*Commentaria, in Isagogen Porphyrii et in omnes libros Aristotelis de dialectica, olim...consilio et...sumptibus facultatis artium in... Academia Lovaniensi, per ...peritissimos viros composita* (Lovanii, 1568)). As we saw, one source for such confusion between methods of teaching and methods of discovery is the ambiguity of Galen's three methods and of Averroës' discussion of *via doctrina* in his explanation of the order in which Aristotle had presented the *Physics*. Digby appears to be unaware of the clarification of these issues by Zabarella and other late-sixteenth-century commentators. The first English commentaries on the *Posterior Analytics* to show the influence of Zabarella's works were not published until after Digby wrote. See G. Powel, *Analysis Analyticorum Posteriorum sive librorum Aristotelis de Demonstratione* (Oxoniae, 1594); J. Case, *Summa veterum interpretum in universam dialecticam Aristotelis* (Oxoniae, 1584). Both cite Zabarella extensively and follow him closely for their interpretation. See Kosman, op. cit., 'The scholastic tradition in Bacon's England', pp. 155–71, especially p. 164.

[1] See note above, p. 60.

[2] *Admonitio*, pp. 20–22; *Pro Mildapetti*, p. 23.

considered superior to, or prior to the tree it produces, simply because it preceded it in time.[1]

The second point which Temple makes against Digby is that he has extended the application of the term *methodus* so far that it is made meaningless. Digby uses the term *methodus* for both invention and judgment in dialectic, whereas Ramus restricted the term's application to the part of judgment concerned with ordering an extended discourse. Digby distinguishes as separate *methodi* applications of a single technique to different types of subject matter and disposition of material derived in differing ways. And in particular, he calls procedures for the discovery of knowledge *methodi*. For Ramists, Temple emphatically states, *methodus* is a term only to be used in the context of teaching and laying out material for clarity and comprehensiveness.[2] No matter how knowledge is acquired (and it certainly arises from knowledge of particulars), its disposition for teaching is always from most general to less general.[3] Hence the divisive technique, starting from the most general precept of an art, is indeed the one and only *methodus*.

Temple's most damaging point against Digby is concerned with the procedure for coming to know discussed by Digby in *Theoria* and touched on briefly in the *De duplici methodo*, and designated by him a kind of method. Digby implies, according to Temple, that there are two procedures to account for the acquisition of new knowledge. The novice uses a process of composition from particular experiences (and Digby uses this as his justification of composition as a teaching method); the privileged few (Digby amongst them) aspire to the 'marvellous'

[1] *Pro Mildapetti*, p. 102f.: 'Id itaque *notius* appello, quod explicari intelligique possit, tametsi illud, quocum splendore notitiae comparatur, minime explicetur et intelligatur...Neque vero si origo notitiae ex observatione rerum singularium sensu preceptarum existat, idcirco res singulares sunt nobis notiores. Vides e minutissimis seminibus originem magnitudinis suae truncum accepisse, semen tamen trunco maius non est...sic etiamsi notitiae in generalibus origo ex observatione singularium defluxit: generalia tamen singularibus luce claritatis praestant. Illud itaque teneatur, methodum non ab ignotioribus ad notiora, aut a tenebris (ut ait Diplodophilus) ad lucem progredi, sed a rebus absolute clarioribus ad obscuriora semper procedere. Assumptionem vero videamus *universalia sunt absolute notiora specialibus*.'

[2] *Pro Mildapetti*, p. 54: 'Non enim methodi est aut conclusionem e generalibus principiis demonstrare, aut factam demonstrationem retexere, aut integrum in partes distinguere, aut finem in media resolvere. Haec non methodi praecepto, sed alterius regulae subsidio fiunt. Non illa specialium inductionem attingit, non in enuntiati et syllogismi possessiones irruit: non specialia generalibus praeponit: non vult denique sese nomine definitivae aut divisivae appellari, cum nihil vel definiat aliquando vel distribuat.'

[3] *Admonitio*, p. 72.

analytic meditative method for discovering forms. Temple both jibes at the elitism of this account, and states categorically that Digby is wrong. All knowledge of general precepts is formed by observation and induction.[1] This Aristotle himself made perfectly clear.[2] In the *Pro Mildapetti* Temple takes all those passages in Aristotle's works used by Digby in support of his methods of coming to know (in particular the critical passages in the second book of the *Posterior Analytics*) and shows accurately how these in fact, in context, refer to or presuppose a process of observation and induction.[3] Since Temple claims only to be concerned with teaching method, he is spared the job of actually giving details of such an induction.

Temple follows Digby in devoting what to us appears an unreasonable amount of space in his two pamphlets to scurrilous comment on his opponent, but the main points of his counter-attack are nevertheless clearly and convincingly made. He explicitly separates out teaching method from methods of discovery in the sciences. He argues persuasively for a single order of intellectual priority in place of the overcomplex scholastic discussion. And he rejects all accounts of coming to know except an induction based on particular sense experiences.

The parallel between Temple's polemic against Digby and Bacon's polemic against the dialectical tradition as a whole is clear. Like Bacon, Temple insists on a careful distinction between discussions of acquisition of knowledge and discussion of presentation of knowledge. And like Bacon, Temple insists on the sovereignty of induction in the discovery of knowledge. But

[1] *Admonitio*, p. 72: 'Agnosco e sensuum observatione et rerum inductione omnem cognitionem defluxisse.' See also p. 74.

[2] *Admonitio*, p. 76: 'Eadem est Aristotelis de artium inventione sententia: a quo certe homine tua jamdudum vehementer aberrat oratio. Ille in posterioribus Analyticis observationem, experientiam, inductionem quod ab hisce omnium artium instituta procreentur, copiose magnificeque celebrat.'

[3] *Pro Mildapetti*, p. 48ff. See in particular Temple's comments on passages from the *Posterior Analytics*, pp. 55–65. e.g.:

'Cap Exod: Fieri non potest ut universalia percipiantur nisi per inductionem. Ergo.
Exod: cap: Neque enim ex universalibus scientia absque inductione, neque per inductionem sine vi sentiendi. Ergo.

Aristoteles quem prius contexuit soritem, eundem deorsum versus retexit. Scientia (inquit) non existit absque universali nec universale efficitur sine inductione, nec inductio efformatur nisi accesserit sensus. Quamdiu nos inductione ista eludes? Quamdiu in scientiae investigandae praecepto delirabis, de inveniendis per inductionem generalibus confitemur: unica tamen est rerum generalium, quae inductione specialium constitutae sunt, ordine collocandarum methodus.' (p. 63).

here the parallel ends. Temple's aim is simply to establish the unique efficacy of a simplified and rationalised version of Ramus' divisive method as a teaching method. Bacon's polemics are a prelude to more grandiose schemes. For him no existing method has any special prerogative as a teaching method. And for him all existing methods, including the existing forms of induction, are inadequate for the acquisition of new knowledge, so that something quite new is called for.

3

Bacon's response to the dialectical tradition

There is no question of tracing Bacon's views directly to particular dialectic texts of the sort discussed in the previous chapter. Bacon rarely cited what he had read, nor have very many books known to have belonged to him survived.[1] It is likely that he came to dialectic, and Aristotle's Organon, by way of the dialectic handbook of Caesarius or Seton (and he may indeed, to judge from other students of the period, have used both). What evidence we have on the subject of university teaching in the 1570s at Cambridge also makes it clear that the study of dialectic occupied the major part of the student's time during the first two years of the four-year BA.[2] Bacon comments, in the *De Augmentis*, on the current practice of introducing students to logic at an early stage in the university course (the average age of entry at Cambridge was fifteen):

It is a general custom (and yet I hold it to be an error) that scholars come too soon and too unripe to the study of logic and rhetoric [at university], arts fitter for graduates than children and novices; for these two rightly taken are the gravest of sciences, being the arts of arts, the one for judgment, the other for ornament; besides they give the rule and direction how both to set forth and illustrate the subject matter. ...And...the premature and untimely learning of these arts has drawn on, by consequence, the superficial and unprofitable teaching and

[1] Bacon left all his books to John Constable, one of his executors, but in view of the state of Bacon's financial affairs at his death the books may well have been seized by creditors. Certainly his library was dispersed. Gorhambury House (the family home) contains nothing of interest (I doubt whether a single book there now belonged to Bacon); the Gray's Inn library likewise yields nothing. The Francis Bacon Library, Claremont, California, owns one book (Diogenes Laertius' *Lives of the Philosophers*) with Bacon's signature. The Folger collection has a small number of such books, which I have not seen.

[2] See L. A. Jardine, 'The place of dialectic teaching in sixteenth century Cambridge', *Studies in the Renaissance* (1974).

handling of them, – a manner of teaching suited to the capacity of children. [IV, 288]

However, in the light of the dialectical background which I have sketched in the last chapter, we are apparently in a position to judge at what some of his remarks on logic and methodology are directed. It is a considerable help in considering Bacon's attitude towards induction and first principles to know that his criticisms of existing treatments of induction, and of the unquestioning acceptance of fundamental principles, may plausibly be read as a reaction to what he learnt from Melanchthon, Caesarius and Seton (or some combination of these) as a student. This is the sort of level at which I am interested in tracing Bacon's attitudes to prevalent ideas in the areas in which he worked.

To make this point clearer, here is a specific example of the approach I have in mind. Bacon rejected Ramus' 'one and only method' as a natural method of presentation:

And first for the 'one and only method,' with its distribution of everything into two members, it is needless to speak of it; for it was a kind of cloud that overshadowed knowledge for awhile and blew over: a thing no doubt both very weak in itself and very injurious to the sciences. For while these men press matters by the laws of their method, and when a thing does not aptly fall into those dichotomies, either pass it by or force it out of its natural shape, the effect of their proceeding is this, – the kernels and grains of the sciences leap out, and they are left with nothing in their grasp but the dry and barren husks. And therefore this kind of method produces empty abridgments, and destroys the solid substance of knowledge. [IV, 448–9]

Bacon rarely refers directly to another author, but in another passage he makes explicit reference to Ramus, and commends him in a limited way for adapting Aristotle's three rules, as a guide to the scope and type of proposition appropriate to any specified field of discussion:

And herein [in determining the *type* of proposition in formulating an art] Ramus merited better in reviving those excellent rules of propositions (that they should be true, universally, primarily, and essentially), than he did in introducing his uniform method and dichotomies; and yet it comes ever to pass, I know not how, that in human affairs (according to the common fiction of the poets) 'the most precious things have the most pernicious keepers.' Certainly the attempt of Ramus to amend propositions drove him upon those epitomes and shallows of knowledge...Nevertheless I must confess that the intention of Ramus in this was excellent. [IV, 453][1]

These two passages make Ramus a *source* for Bacon's views only in a limited sense. They suggest that Bacon takes familiarity with the controversial issues in Ramus' dialectic (the three rules and the dichotomous method) for granted as background to his own discussion. And they show that as far as Bacon himself is concerned, he recognises that these issues belong strictly in the realm of *presentation* of knowledge, and judges them accordingly. As I shall show in chapter 6 Bacon uses Aristotle's three rules himself, in framing the conditions of 'certainty' and 'freedom' which his own scientific propositions must fulfil. That is to say, in the conventional manner of Aristotle himself (or as discussed under 'demonstration' by a dialectician like Caesarius), which owes nothing to Ramus' rather eccentric reading of Aristotle.

It is tempting to see direct influence of William Temple on Bacon. We have noted Temple's unusual clarity about the priority of induction in the process of cognition and the distinction between acquiring knowledge and teaching, and these are key features of Bacon's approach to knowledge.[2] On intellectual grounds this is more plausible than the long-standing suggestion that Digby taught and influenced Bacon.[3] Either way, Bacon had certainly left Cambridge some years before the circulation there of Digby and Temple's polemical pamphlets, having completed two years of the four-year BA (that is, the preliminary training, almost exclusively in dialectic, to reach 'sophister' level).[4] Once

1 For Ramus' treatment of the three rules, which Bacon summarises quite accurately, see above, p. 52. See also [III, 236] where Bacon refers obliquely to Ramist adaptation of the rules.

2 It is probable that Bacon knew Temple. Temple became secretary to Essex in 1594; Bacon was already a close friend of Essex in 1591, and his brother Anthony was also a secretary to Essex. There is a reference in Bacon's correspondence to his attempting to secure a knighthood for a William Temple in 1607–8 (Temple was in fact knighted a few years later) [X, 2–3].

3 Digby's direct influence was suggested by J. Freudenthal, 'Beiträge zur Geschichte der englischen Philosophie [I]', *Archiv für Geschichte der Philosophie*, 4 (1891), pp. 450–77, and has since become standard in the literature. In a recent book A. Crescini devotes nearly 100 pages to Baconian method, starting from the assumption that Digby exerted direct and lasting influence on Bacon's philosophical outlook. See *Il problema metodologico alle origini della scienza moderna* (Rome, 1972), pp. 87–181.

4 Bacon attended Cambridge from 5 April 1573 until Christmas 1575 (with a short break because of the plague). He was at Trinity, under the direct surveillance of Whitgift (the

again, I suggest only that the Digby/Temple dispute typifies contemporary attitudes and opinions, particularly in the context of the dialectic teaching programme at Cambridge, in which both protagonists participated.[1]

The main themes of Bacon's discussion of his new logic or *Novum Organum* place him squarely within the sixteenth-century dialectical tradition. First and foremost, he distinguishes rigorously between a logic for the discovery of new knowledge and one adequate for the presentation of an existing body of knowledge, and lays the blame for confusion of the two at the door of the 'dialectici':

> Invention is of two kinds, very different; the one of arts and sciences, and the other of speech and arguments [premises]. The former of these I report altogether deficient...For in the first place, Logic [Dialectic] says nothing, no nor takes any thought, about the invention of arts...but passes on, merely telling men by the way that for the principles of each art they must consult the professor of it [cuique in sua arte credant]. [IV, 407–8]

> The invention of arguments is not properly an invention; for to invent is to discover that we know not, not to recover or resummon that which we already know. Now the use and office of this invention is no other than out of the mass of knowledge which is collected and laid up in the mind to draw forth readily that which may be pertinent to the matter or question which is under consideration. For to him who has little or no knowledge on the subject proposed, places of invention [Loci Inventionis] are of no service...not to be nice about words, let it be clearly understood, that the scope and end of this [topical] invention is readiness and present use of our knowledge, rather than addition or amplification thereof. [IV, 421–2]

master). In 1576 Bacon went to France, and was at the French court in the train of the French Ambassador to Elizabeth's court, Sir Amias Paulet, until his father's death in 1579, when he returned to Gray's Inn (where he had been enrolled in 1576) and took up law as a profession. Kearney suggests that this is a typical educational pattern for a young gentleman destined for public service. See H. Kearney, *Scholars and Gentlemen: Universities and Society in Pre-Industrial Britain 1500–1700* (London, 1970), pp. 25–8. See also W. R. Prest, *The Inns of Court under Elizabeth I and the Early Stuarts 1590–1640* (London, 1972). Temple was in Paris in 1579, see *Admonitio*, p. 19.

[1] For general background on the English universities see *Rashdall's Mediaeval Universities*, ed. Powicke and Emden (Oxford, 1936), vol. III. On renaissance education in general see Foster Watson, *The English Grammar Schools to 1660: their Curriculum and Practice* (Cambridge, 1908); T. W. Baldwin, *William Shakspere's small Latine and lesse Greeke* (Urbana, 1944). For a contemporary account of the schools curriculum in the early seventeenth century see C. Hoole, *A new discovery of the olde arte of teaching schoole* (London, 1660). On social background see M. H. Curtis, *Oxford and Cambridge in Transition 1558–1642* (Oxford, 1959); J. Simon, *Education and Society in Tudor England* (Cambridge, 1965); Kearney, op. cit.

For the end which this science of mine proposes is the invention not of arguments but of arts; not of things in accordance with principles [principiis consentanea], but of principles themselves; not of probable reasons [rationes probabiles], but of designations and directions for works. [IV, 24]

Here not only the attitudes criticised, but the very phrases in which these are expressed, are drawn from the dialectic handbook, as comparison with my first chapter will readily show.

In maintaining that for the purpose of discovery of knowledge the alternative to such dialectical invention is an *induction* Bacon equally conforms with the parenthetical remarks on *principia* of the dialecticians.[1] But he heaps scorn on dialecticians (like Seton) who confound dialectical induction by incomplete enumeration with this induction of concepts:

[the lack of an adequate logic of discovery] is demonstrated (if you observe it carefully) by the form of induction which Logic [Dialectica] proposes, as that whereby the principles of sciences may be invented and proved [probentur];[2] which form is utterly vicious and incompetent, and so far from perfecting nature, that contrariwise it perverts and distorts her. For...the mind does of herself by nature manage and act an induction much better than logicians [dialectici] describe it; for to conclude upon a bare enumeration of particulars (as the logicians do) without instance contradictory, is a vicious conclusion; nor does this kind of induction produce more than a probable conjecture. [IV, 410]

But the induction which is to be available for the discovery and demonstration of sciences and arts, must analyse nature by proper rejections and exclusions; and then, after a sufficient number of negatives, come to a conclusion on the affirmative instances: which has not yet been done or even attempted, save only by Plato, who does indeed employ this form of induction to a certain extent for the purpose of discussing definitions and ideas. [IV, 97–8]

(As we saw, even the remark commending Plato's handling of definitions is a feature of dialectic handbooks.)

The inductive method which Bacon proposes as his own method of discovery is supposed to be a natural method. It corresponds stage by stage to the process of perception of natural phenomena. Primitive perceptions are recorded, sifted

[1] Compare, for instance, Bacon's remarks with Caesarius, *Dialectica* (1568 ed.), fo. T1v.

[2] Compare the phrasing here with Seton's account of the induction which supplies confirmation of principles: 'Per inductionem utcunque probari possunt scientiarum principia, quae nullo modo demonstrari queant', *Dialectica*, ed. cit., fo. Ov.

and tabulated under their most evident groups, and then an eliminatory induction is carried out. These three stages in the 'interpretation of nature' supposedly correspond to the natural functions of the senses, memory and reason. By employing these faculties appropriately the inductive method is supposed to yield fundamental principles (the basic generalisations on which all human decisions and actions depend) free from distortion, and which hold under all possible circumstances (are *certain* rather than *probable*).[1] As we saw in the last chapter, Ramus claimed his dialectic as a 'natural' method: the rules of dialectic are the rules governing the natural ways in which the mind sifts and classifies. Bacon denies that *any* logical system which comes into operation (as Ramus' does) after the fundamental terms and principles have already been established can hope to remedy the errors which are '(as the physicians say) in the first digestion; which is not to be rectified by the subsequent functions' [IV, 411]. And he explicitly rejects the belief (crucial for Ramus and all Platonists) that any dialectical, classificatory method can accurately reproduce the hierarchy of genera and species of nature [III, 553]. For Bacon no method of presentation, no essentially *mnemonic* method, can ensure such fidelity to nature; hence his rejection both of Ramist method, and of Lullian method as impostures [IV, 454; IV, 448].[2] It is on principles themselves that the first assault must be made, and this must be done by means of a method whose every stage mimics the operation on

[1] See for instance [III, 552–3; IV, 19; IV, 127].

[2] As we see here, it is entirely characteristic of Bacon's approach to knowledge that he should deny that memory is itself capable of organising its store of material internally to correspond to the natural ordering of things in the universe. He points out that it is hardly to be expected (given the distortions of our unaided perception) that any classification which the memory yields will correspond to the true divisions of nature [III, 553]. At best it will organise a store of sense impressions for internal consistency ('ex analogia hominis'), whereas science must organise such a store as it reflects things in nature directly ('ex analogia universi'). The view that the 'art of memory' or artificial procedures for retrieving the contents of the memory store can yield scientific knowledge immediately was, however, one which was widely held in the period. See P. Rossi, *Clavis Universalis* (Milan, 1960); F. Yates, *The Art of Memory* (London, 1966), for a full account of the various mnemonic techniques proposed in the period as the basis for complete understanding of the natural world. Ramus, for instance, believed that memory is 'a sort of shadow of disposition'; that is, he believed that the process whereby terms and propositions formed in the mind are arranged in discourse is a systematisation of the natural activity of memory (see Rossi, op. cit., p. 140). Ramon Lull's 'method' is an early attempt at generating the whole of natural knowledge from the memory store. Bacon called Lull's method 'a method of imposture', 'being nothing but a mass and heap of the terms of all arts, to the end that they who are ready with the terms may be thought to understand the arts themselves' [IV, 454]. For Bacon such methods have a place only in oratory, where they facilitate retrieval of appropriate material [IV, 435–6].

immediate sense perceptions of a faculty of the mind, and corrects its inherent and acquired defects. Thus his own 'ministrations to the memory', with which the interpretation of nature begins, are external aids which compensate for the restricted natural capacities of memory [III, 552–3; IV, 127]. Unlike mnemonic methods, which reinforce the natural bent of the memory without regard for the secure foundation for the material manipulated, this series of operations compensates in each of its stages for a source of error in *acquisition* of material of unaided memory.

An essential feature of Bacon's system is his optimistic view of the objectivity of sense experience. The senses do not obstruct our view of nature entirely; they are limited in their capacities and tend to mislead, but these limitations can be overcome. The investigator has so to frame his investigation that the best possible use is made of the senses as they are:

> Now the senses, though they often deceive us or fail us, may nevertheless, with diligent assistance, suffice for knowledge; and that by the help not so much of instruments (though these too are of some use) as of those experiments which produce and urge things which are too subtle for the sense to some effect comprehensible by the sense.
> [IV, 412][1]

Although Bacon approved the caution of the sceptics in questioning the foundations of all human knowledge, he explicitly dissociated his own approach to knowledge from theirs:

> The doctrine of those who have denied that certainty could be attained at all, has some agreement with my way of proceeding at the first setting out; but they end in being infinitely separated and opposed. For the holders of that doctrine assert simply that nothing can be known; I also assert that not much can be known in nature by the way which is now in use. But then they go on to destroy the authority of the senses and understanding; whereas I proceed to devise and supply helps for the same. [IV, 53]

> It will also be thought that by forbidding men to pronounce and set down principles as established until they have duly arrived through the intermediate steps at the highest generalities, I maintain a sort of suspension of the judgment, and bring it to what the Greeks call *Acatalepsia*, – a denial of the capacity of the mind to comprehend truth. But in reality that which I meditate and propound is...not denial of the

[1] See also [IV, 26, 58; III, 244].

capacity to understand, but provision for understanding truly; for I do not take away authority from the senses, but supply them with helps; I do not slight the understanding, but govern it. [IV, 111–12][1]

It is interesting to note that Agricola, in his *De Inventione*, shows a marked sympathy towards Ciceronian scepticism,[2] and that this is linked with his belief that eloquence will yield humanity quite as much knowledge as technical quibbling and demonstration. Here again, Bacon's whole outlook is at odds with the 'literary' approach of the dialecticians. Certainty is for Bacon an attainable goal for knowledge, if the inductive method is substituted for traditional 'false forms of demonstration and ill-ordered methods of reasoning'.[3]

As one would expect, Bacon, as a professional speech maker and a writer, showed a lively interest in method of presentation for its own sake. And as one might also expect, his clear cut separation of a 'natural' method for deriving first principles from all the various methods discussed by the dialecticians affected his attitude towards the possibilities for using some sort of methodical presentation didactically and persuasively.

In the course of his university training in the liberal arts, Bacon would have encountered discussion of disposition or presentation of material at two points. One of these was in the course of his dialectic training, as I have outlined in chapter one. The other was during his grounding in *rhetoric*. Since sixteenth-century dialectic was already heavily literary in its emphasis, there was in fact very little need to supplement this course (which, as I pointed out, apparently occupied much of the student's time during his first two years at Cambridge) with a detailed theoretical treatment of rhetoric. Caesarius' *Rhetorica*, for example, is openly presented as subsidiary to his dialectic manual, and effectively contains material (much of it practical instruction in delivery and selection of material) from Cicero, Quintilian and the *Ad Herennium* which of its nature falls outside the terms of reference of the dialectic manual. Much of the student's actual rhetoric training probably consisted in analy-

[1] See also [IV, 69; IV, 411–12].

[2] On Ciceronian scepticism in the period, see C. B. Schmitt, *Cicero Scepticus: a study of the influence of the Academica in the Renaissance* (The Hague, 1972).

[3] See R. E. Snow, *The Problem of certainty: Bacon, Descartes and Pascal* (unpublished Ph.D. diss., Indiana, 1967), pp. 9–61.

sis of passages from Cicero's orations or Ovid (both set texts in the period in Cambridge), and in prose and verse composition. However, the rhetoric manual did contain a brief treatment of *disposition* of whole narratives or orations. This relies heavily on a single passage in the *Ad Herennium*, and consists of some brief remarks on the various 'orders of priority' which may be observed in setting out a narrative: the order of the 'parts of the oration' (introduction, thesis, arguments for and against the question, summing up, conclusion), temporal ordering for historical events, and so on.

One of Ramus' central claims was that his dichotomous method could be substituted for rhetorical disposition, as well as providing a unique teaching method in dialectic. In the single study of the 'ars disserendi', the general art of discourse, a single study of disposition is appropriate to all types of expression. On the same grounds, Agricola brought together rhetorical and dialectical disposition under one heading, linking the discussion of the *Ad Herennium* with discussions of syllogism and teaching method. Like them, Bacon discusses all methods of disposition together as the *Wisdom of Transmission*:

> Let us now come to the doctrine concerning the Method of Discourse. This has been commonly handled as a part of Logic [dialectica]; and it also finds a place in Rhetoric, under the name of *Disposition*. But the placing of it in the train of other arts has led to the passing over of many things relating to it which it is useful to know. I have therefore thought fit to make the doctrine concerning Method [Methodus] a substantive and principal doctrine, under the general name of *Wisdom of Transmission* [Prudentia Traditivae]. [IV, 448]

For Bacon as for Agricola, method of transmission is a question of *prudentia* rather than *sapientia*. That is, it is a practical, *ad hoc* study, depending in every case on a combination of circumstances: the accessibility of the material to be transmitted, the state of mind of the audience, the desired effect of communication, and so on. Any available means may be resorted to in presenting material in order to achieve the desired response. That is, any procedure which has proved successful in the past, or simply exists as a literary form, may serve to transmit a body of information, to make it palatable and easy to digest. Only one method has any privileged status for Bacon: scientific principles communicated by the stages of the inductive method itself will be perspicuous and unmisleading, because it is open to the listener to retrace in its entirety an infallible method of discovery [IV, 449]. Otherwise all means of presentation exploit to a greater or lesser extent the prejudices and credulity of an audience.

Having rejected the idea that any *methodus sermonis* except the inductive method itself is a 'natural' method of presentation (that is, one which presents the material without distortion, as it would be acquired in nature), Bacon includes as *methodi* within the *Wisdom of Transmission* a range of didactic and literary presentation techniques beyond anything discussed in Agricola. And he sets traditional didactic methods (the question and answer procedure of the scholastic manual, and the three Galenic methods amongst them), together with Ramist method, on a par with other persuasive techniques which are appealing or convenient for an audience, but at the cost of distorting the subject matter.

We may characterise Bacon's attitude towards presentation of existing knowledge (the focus of humanist dialectic) as follows. He abandons the distinction between what traditionally had been designated dialectical and rhetorical disposition, as Agricola had done. But where Agricola claimed that removing such a distinction means that all literature is 'teaching', Bacon claims the opposite. For him all teaching is 'insinuative'; all presentation is misrepresentation to some specified end. He therefore lumps together as techniques for persuading rather than explaining, dialectical invention [IV, 421–3], disputational syllogistic [IV, 49], even the collection of principles and beliefs which have traditionally served as the premises for disputation and ground of the arts [IV, 42]. These belong for Bacon with the tropes and schemes of the rhetoric manual, and the strategies for persuasion in a style-book. All methods of presentation except straightforward recapitulation of the process of discovery are 'rhetorical' in our modern loose sense of the word. The range of methods which insinuate their conclusions runs from the schoolmaster's epitome to the lawyer's heavily embellished forensic syllogisms. Any type of presentation may be selected to suit the occasion and the audience; any literary form (a poem, a parable) may be chosen to set off the material to its best advantage.

4

Bacon's theory of knowledge

The traditional Aristotelian view of the most fundamental principles of the sciences was that they should be fixed and certain, prior and better known in nature, and at once definitions and causes.[1] Bacon takes it for granted that the goal of science is truths which fulfil these criteria; truths which cannot be other than they are and are prior in the order of nature.

In any subject the most general definitions are prior in the order of nature. Thus Bacon's goal in science is the most general definitions of natural philosophy. And like such commentators as Zabarella, Bacon appears to take the 'realist' view that causes and principles which are better known in nature are those which describe the true structure of nature which underlies the processes apparent to us. Although it has been shown that such scholastic writers as Niphus were aware of the conjectural nature of the principles used to provide the premises for syllogistic explanation of effects,[2] Aristotelian commentators were on the whole opposed to the non-realist view (expounded with great subtlety and insight by Ockham)[3] that the most that can be achieved in science is to 'save the phenomena' by positing hypotheses which are as economical and convenient for predictive purposes as possible. Thus, whilst Bacon is anti-Aristotelian in his belief that celestial phenomena are subject to the same principles as terrestrial phenomena, his grounds for rejection of the Copernican system show an essentially Aristotelian view of

[1] See e.g. Melanchthon, *Erotemata Dialectices* (Witebergae, 1562 ed.), fo. R2r-v; Caesarius, *Dialectica* (Coloniae Agrippinae, 1568 ed.), fo. 53v.

[2] See L. A. Kosman, *The Aristotelian Backgrounds of Bacon's 'Novum Organum'* (unpublished Ph.D. diss., Harvard, 1964), pp. 116–38, 'Aristotelian theory of science in the Renaissance: Augustino Nifo and Jacopo Zabarella'.

[3] Kosman, op. cit., pp. 101–15.

the nature of natural philosophy.[1] For him such 'hypotheses' could not provide principles with natural priority. Hence Bacon rejects the Copernican system in astronomy, which he regards as merely a hypothesis consistent with the available data, and useful for making astronomical predictions, but in no way guaranteed as true:

In like manner astronomy presents only the exterior of the heavenly bodies (I mean the number of the stars, their positions, motions, and periods), as it were the hide of the heavens; beautiful indeed and skilfully arranged into systems; but the interior (namely the physical reasons) is wanting, out of which (with the help of astronomical hypotheses) a theory might be devised which would not merely satisfy the phenomena (of which kind many might with a little ingenuity be contrived), but which would set forth the substance, motion, and influence of the heavenly bodies as they really are...And it is the absurdity of these opinions [astronomical hypotheses] that has driven men to the diurnal motion of the earth; which I am convinced is most false...all labour is spent in mathematical observations and demonstrations. Such demonstrations however only show how all these things may be ingeniously made out and disentangled, not how they may truly subsist in nature; and indicate the apparent motions only, and a system of machinery arbitrarily devised and arranged to produce them, – not the very causes and truth of things. [IV, 348–9]

As for the hypotheses of astronomers, it is useless to refute them, because they are not themselves asserted as true, and they may be various and contrary one to the other, yet so as equally to save and adjust the phenomena. [V, 557]

Bacon's terminology in outlining the goals of his inductive method shows his commitment to an Aristotelian framework for sciences. The following passages should be compared with my earlier discussion of the Aristoteliain attitudes of the dialectic handbook (chapter 1, p. 34):

Lastly, the true Form is such that it deduces the given nature from some source of being which is inherent in more natures, and which is better known to nature [quae inest pluribus et notior est naturae (ut loquun-

[1] See W. H. Donahue, *The Dissolution of the Celestial Spheres* (unpublished Ph.D. diss., Cambridge, 1972) for a comprehensive survey of the astronomical beliefs of Bacon's contemporaries. R. N. Blake, 'Theory of Hypothesis among Renaissance Astronomers', in Madden, Blake and Ducasse, *Theories of Scientific Method* (Washington, 1960), pp. 22–49, and Kosman, op. cit., pp. 139–47, review renaissance attitudes to hypothesis in astronomy, but are slightly misleading because they fail to point out the traditional gulf between astronomy, a practical art linked to mathematics, and natural philosophy in which certain knowledge is attainable.

tur)] than the Form itself. For a true and perfect axiom[1] of knowledge then the direction will be, *that another nature be discovered which is convertible with the given nature, and yet is a limitation of a better known nature, as of a true and real genus* [*ut inveniatur natura alia, quae sit cum natura data convertibilis, et tamen sit limitatio naturae notioris, instar generis veri*]. [IV, 121–2]

Now my plan is to proceed regularly and gradually from one axiom[1] to another, so that the most general are not reached till the last: but then when you do come to them you find them to be not empty notions, but well defined, and such as nature would really recognise as better known to her [talia quae natura ut revera sibi notiora agnoscat], and such as lie at the heart and marrow of things. [IV, 25]

[The old method] begins at once by establishing certain abstract and useless generalities, [the inductive method] rises by gradual steps to that which is prior and better known in the order of nature [quae revera naturae sunt notiora]. [IV, 50]

Another characteristic feature of the first principles envisaged by Bacon for science is that they are to be definitions in terms of the essential qualities of natural phenomena. This automatically means that mathematics plays only a subsidiary rôle in Baconian science. Number and quantity are, in the Aristotelian scheme, accidental, not essential attributes of bodies. Hence quantitative description of natural processes is at least in theory subsidiary to their explanation in terms of qualitative causes. Indeed, Bacon

[1] Bacon uses the term '*axioma*' to mean any general proposition, not simply the first principles of knowledge. In this he follows Ramus. Ellis comments: 'Hasse, an early commentator on Ramus, remarks that the word is used in the same way by Cicero, who probably took it from the Stoics' [I, 136]. R. Gocklenius, a follower of Ramus, in his *Lexicon Philosophicum, quo tanquam clave philosophiae fores aperiuntur* (Francofurti, 1613), begins his entry under *axioma* (p. 135): 'Primo Axioma Logicis Stoicis, Ciceroni et Rameis significat Enunciatum seu sententiam seu Propositionem, hoc est compositionem notionum, qua aliquid affirmatur aut negatur.' He then gives as its secondary meaning: 'Secundo Peripateticis significat πρότασιν indemonstrabilem.' Temple justifies this use of *axioma* interestingly as follows: 'Quod attinet ad vocem hanc (Axioma) notat illa quidem huiusmodo dispositionem quae animo et mente interius efficitur...Nam dialectica est ars rationis et intelligentiae non linguae et externi sermonis: adeo ut qui solus secum commentatur et nunquam eloquitur, is possit ratione uti et artem dialecticam exercere. Ac proinde cum illa vox (Axioma) dispositionem argumenti cum argumento ab animo interius factam designet: caeterae autem voces (nempe enuntiatum, pronuntiato, effatum) potius exteriorem illam dispositionem, quae lingua et caeteris in ore instrumentis efficitur, significatione attingant: vocabulum Axiomatis est caeteris praeferendum. Nam de nomine Propositionis quid dicam? Prima pars plenae comparationis appellatur propositio: prima pars syllogismi nominatur propositio: quod si etiam dispositio illa, qua unum argumentum enuntiatur de altero, diceretur propositio, quanta esset huius vocis ambiguitas et confusio? Retineatur ergo nomen Axiomatis.' Temple, *Dialectica* (Cantabrigiae, 1584), pp. 66–7. On Bacon's terminology in general see R. Eucken, *Geschichte der philosophischen Terminologie* (Hildesheim, 1964), pp. 84–5.

believed that it was extremely misguided to search for laws of proportionality in nature:

For men believe that if the quantity be increased or multiplied, the power and virtue is increased or multiplied proportionately. And this they postulate and suppose as if it had a kind of mathematical certainty; which is utterly false. A leaden ball of a pound weight dropped from a tower reaches the ground in (say) ten seconds: will a ball of two pounds weight (in which the force of natural motion, as they call it, ought to be doubled) reach the ground in five seconds? No, but it will take almost the same time in falling, and will not be accelerated in proportion to the increase of quantity. Again, suppose one drachm of sulphur mixed with half a pound of steel will melt it and make it liquid; will therefore one ounce of sulphur mixed with four pounds of steel be able to melt it? This does not follow; for it is certain that the obstinacy of matter in the patient is more increased by quantity than the active power of the agent... Men should therefore consider the story of the woman in Aesop, who expected that with a double measure of barley her hen would lay two eggs a day; whereas the hen grew fat and laid none. [IV, 414–5]

What Bacon objected to in the Aristotelian account was not the general theory of knowledge which it implied but the naivety of the means by which it supposed such knowledge to be discovered. For Bacon simple syllogistic reasoning of the kind used in demonstration is a totally inadequate tool for getting at the underlying causes of natural phenomena. But unlike Agrippa, Sanchez and Montaigne[1] he does not consider the shortcomings of the senses and the inadequacy of the traditional syllogistic methods of discovery to be grounds for scepticism. The induction which Bacon proposes to substitute for the Aristotelian methods of discovery of first principles is supposed to guarantee the certainty, or natural priority of first principles by ensuring that at every stage, beginning with basic sense perception, the conjectural element in traditional methods is eliminated by appeal to experience. Only in this way can principles which are 'grounded in matter' be attained.

According to Bacon, although both our senses and reasoning faculty mislead us in our unguided observations of natural phenomena, certain knowledge is not thereby put beyond our

[1] See R. H. Popkin, *The History of Scepticism from Erasmus to Descartes* (Assen, 1960), pp. 17–43 'The revival of Greek Scepticism in the Sixteenth Century', and pp. 44–65 'Michel de Montaigne and the "Nouveaux Pyrrhoniens"'; H. Van Leeuwen, *The problem of certainty in English thought, 1630–90* (The Hague, 1963).

grasp. With careful guidance, and compensating techniques for the failings of the senses, certainty can be attained.

The starting point for Bacon's procedure for arriving at certain truths in nature is a survey of the inherent and acquired defects of the senses and the mind, which such a procedure must circumvent. There are some predispositions of the mind, according to Bacon, which impede progress towards certain definitions. Bacon called these 'Idols' – Idols of the Tribe, Idols of the Cave, Idols of the Market Place, and Idols of the Theatre.[1] The Idols of the Tribe are errors attributable to characteristics of human nature itself, which taints perceptions, and encourages men to form systems 'ex analogia hominis' rather than 'ex analogia universi' – systems appearing consistent to the human mind, rather than systems consistent with the way things are in nature [IV, 54; I, 163]. The nature of the animal spirit which pervades the human body itself distorts the sensations recorded in the mind from external stimuli [IV, 58–9]. Even the most simple registering in the mind of a sense-impression is produced by a motion of animal spirits [V, 324], which therefore characterises every mental operation:

> And therefore one of the moderns[2] has ingeniously referred all the powers of the soul to motion, and remarked on the conceit and precipitancy of some of the ancients, who in too eagerly fixing their eyes and thoughts on the memory, imagination, and reason, have neglected the Thinking Faculty, which holds first place. For he who remembers or recollects, thinks; he who imagines, thinks; he who reasons, thinks; and in a word the spirit of man, whether prompted by sense or left to itself, whether in the functions of the intellect, or of the will and affections, dances to the tune of the thoughts; and this is the frisking of the Nymphs [i.e. spirits, as Bacon has glossed on the previous page].
> [IV, 325]

Furthermore, a natural function of reason is to generalise from what is observed, to prepare men for the future. In generalising

[1] As Bacon himself notes, his survey of the defects of the mind has much in common with sceptical accounts of the impediments to all certainty. See e.g. Franciscus Sanchez, *De multum nobili et prima universali scientia Quod Nihil Scitur* (1581). Sanchez begins his demolition of certainty as a goal of knowledge by showing that all definitions are mere forms of words (Francofurti, 1618 ed., p. 13); that words are of unstable application and depend upon vulgar usage (p. 15); that the syllogism is inadequate as a tool of inquiry (pp. 20–22); that demonstration cannot yield certain knowledge (pp. 25–30).

[2] The 'modern' in question is Telesius. For a clear account of Telesius' theory of spirits see N. C. van Deusen, *Telesio: The First of the Moderns* (New York, 1932) pp. 53–71.

men tend to exaggerate the extent to which instances resemble one another, to ignore counter instances which do not fit their theories, and to draw all evidence together to support some particularly appealing possibility of order. Whether influenced by the imagination, which can 'feign' order and similarity where none is to be found in nature, or by the affections, which naturally abhor tedious and extended investigations, these errors lead to systematic misrepresentation in the mind of natural data. These are the most primitive and basic impediments to the growth of human knowledge [IV, 55–9].

The Idols of the Cave are errors arising from an individual investigator's personal habits of mind and predilections. By favouring one aspect of interpretation rather than others which are equally probable, or by over-emphasising a single appealing insight, an investigator may misrepresent his material and produce fallacious theories. Such errors can only be avoided by regarding with deep suspicion any interpretation of the data which is immediately and strikingly appealing [IV, 59–60]. A particularly striking example of such personal prejudices is the tendency of some minds to pick out and emphasise the similarities between individual occurrences, and of others to stress the differences between superficially similar instances. The only solution which avoids such distortions is to alternate the two procedures [IV, 60]; Bacon classes himself amongst those who combine the two inclinations, and hence avoid either extreme:

> And for myself, I found that I was constructed more for the contemplation of truth than for anything else, as having a mind agile enough to recognise the resemblance of things (and this is the most important), and sufficiently steadfast and eager to observe the refinements of their diversity. [III, 518]

The Idols of the Market Place are the errors which are automatically introduced into any discussion of natural phenomena if the investigator accepts the received terminology and holds to the basic classifications which are assumed in all application of terms according to common usage. It is extremely difficult for an investigator to reorganise and reclassify against the grain of existing theory, since the common application of terms automatically upholds the prevailing set of opinions.[1]

[1] There has been a prolonged debate about a detail in Bacon's discussion of the misleading power of words. Bacon divides the Idols into *innate* and *adventitious* (acquired) misconcep-

Even definitions, Bacon stresses, involve accepting some terms as given, and may not, therefore, avoid this difficulty [IV, 60–2]. However, at a very rudimentary level Bacon is prepared to accept the possibility of applying general terms to particularly well-marked species. He is prepared to use them as building blocks for definitions which are not formed according to vulgar notions.[1] Bacon believed that his step-by-step procedure for grouping observed instances provides a viable alternative to the over-hasty and distorted concept formation implicit in all ordinary language usage:

> Even definitions cannot cure this evil in dealing with natural and material things; since the definitions themselves consist of words, and those words beget others: so that it is necessary to recur to individual instances, and those in due series and order; as I shall say presently when I come to the method and scheme for the formation of notions and axioms [the inductive method]. [IV, 61]

Finally, the Idols of the Theatre are errors resulting from the adoption of fully-fledged and fallacious systems offered as explanations for natural processes, which persuade by their internal consistency. These are the most easy to eradicate, since they are external in origin and appeal only in the same way that pleasing and consistent fictions in literature are appealing [IV, 62–9].

Bacon's aim in repeatedly characterising for his readers the pitfalls of the various types of Idols[2] is to hammer home to the

tions, and according to some critics he is inconsistent in his allocation of Idols to these two groups. In [III, 396; IV, 61; IV, 433] he appears to treat the Idols of the Market Place as acquired. In [III, 242; III, 245] he also implies that such Idols are innate, since he calls them 'Idols of the Palace', 'Palace' being one of Bacon's metaphors for the human mind [IV, 335] (F. Anderson, *The Philosophy of Francis Bacon* (Chicago, 1948) and P. Rossi suggest that 'Palace' is a misprint for 'Place'!). No real inconsistency is involved. In the first case words are treated as something which is learned or acquired, but which leads to errors which are difficult to eradicate. In the second case they are treated as part of the memory store itself, and the defect is considered to reside in the functioning of memory itself, and hence to be innate. For the debate see A. Levi, *Il pensiero di F. Bacone considerato in relazione con le filosofie della natura del Rinascimento e col razionalismo cartesiano* (Turin, 1925), pp. 321–2; P. Rossi, *F. Bacone: dalla magia alla scienza* (Bari, 1957), transl. Rabinovitch (London, 1968), pp. 160–6; Spedding [I, 113–17].

[1] Bacon believed that the most fundamental and unambiguous of natural kinds can reliably be established by simple, careful observation: 'Our notions of less general species, as Man, Dog, Dove, and of the immediate perceptions of the sense, as Hot, Cold, Black, White, do not materially mislead us; yet even these are sometimes confused by the flux and alteration of matter and the mixing of one thing with another' [IV, 49]. Such basic notions are admissible by virtue of the fact that 'there is in natural objects a promiscuous resemblance one to another' [V, 505]. See also [IV, 292]. In this way Bacon sidesteps the philosophical problem raised by admitting the existence of species as well as individuals as an unproblematic basis for a theory of knowledge.

[2] See [III, 241, 245, 394, 536, 548; IV, 27, 431].

would-be investigator of natural phenomena the extent to which the most obvious procedures of interpretation are mere distortions and prejudices of the human faculties:

As for the detection of False Appearances or Idols, Idols are the deepest fallacies of the human mind. For they do not deceive in particulars, as the others do, by clouding and snaring the judgment; but by a corrupt and ill-ordered predisposition of mind, which as it were perverts and infects all the anticipations of the intellect. For the mind of man (dimmed and clouded as it is by the smooth covering of the body), far from being a smooth, clear, and equal glass (wherein the beams of things reflect according to their true incidence), is rather like an enchanted glass, full of superstition and imposture. [IV, 431]

On the whole Bacon's attitude is that these errors are best avoided, rather than corrected, since the effectiveness of correction cannot be guaranteed. His method aims at by-passing the most obvious sources of distortion in perception and interpretation. The Idols therefore bear a relation to the inductive method analogous to that which cautionary lists of fallacious arguments bear to syllogistic:

The formation of ideas and axioms by true induction is no doubt the proper remedy to be applied for the keeping off and clearing away of idols. To point them out, however, is of great use; for the doctrine of Idols is to the Interpretation of Nature what the doctrine of the refutation of Sophisms is to common Logic. [IV, 54][1]

Armed with a clear understanding of these 'fallacies' of the human faculties, the certainty which is the goal of Baconian science can, according to Bacon, be attained:

Now my method, though hard to practise, is easy to explain; and it is this. I propose to establish progressive stages of certainty. The evidence of the sense, helped and guarded by a certain process of correction, I retain. But the mental operation which follows the act of sense I for the most part reject; and instead of it I open and lay out a new and certain path for the mind to proceed in, starting directly from the simple sensuous perception. [IV, 40]

[1] Caesarius characterises the treatment of logical sophisms in the dialectic handbook as enabling the student to distinguish valid from invalid arguments, and emphasises particularly the *constructive* use of this treatment in pointing out to the student what are the basic errors in arguing. Caesarius compares their rôle in dialectic to that of Donatus' *Barbarismus* in grammar, which gives the student examples of infelicities and errors in Latin style and syntax. Caesarius, *Dialectica* (Coloniae Agrippinae, 1568), fo. Bb4r-Bb4v.

Of the traditional methods for arriving at certain knowledge which make no allowance for the fundamental distortions of the Idols, and hence are themselves irreparably distorting, the one against which Bacon levelled constant criticism was syllogistic. He does not, however, criticise directly the syllogistic methods of demonstration favoured by commentators like Zabarella as techniques for arriving at certain first principles (see chapter 1, pp. 54–8). Indeed, Bacon makes no explicit reference to these sophisticated discussions of scientific demonstration. His criticisms are apparently aimed rather at the orthodox discussions of syllogistic and first principles in the dialectic handbook.[1]

Since logical (dialectical) invention takes the first principles of sciences for granted, all subsequent syllogistic reasoning based upon them is merely probable, without any guarantee of certain truth:

> For logical invention does not discover principles and chief axioms, of which arts are composed, but only such things as appear to be consistent with them. For if you grow more curious and importunate and busy, and question her of probations and invention of principles or primary axioms, her answer is well known: she refers you to the faith you are bound to give to the principles of each separate art. [IV, 80–1][2]

Even if 'sense and experience' could guarantee the truth of the first principles, subsequent syllogistic inference based upon them necessitates the invention of 'middle terms', whose validity also needs to be guaranteed. Once again, this is not discussed by the dialecticians.[3] On two counts, therefore, syllogistic alone is not adequate to ensure the certainty of knowledge:

> And it is not the laborious examination either of consequences of arguments or of the truth of propositions that can correct that error;

[1] From the terminology which Bacon uses in his discussion of the failure of logicians (dialectici) to come to terms with the problems of first principles I suspect that the only discussion of the methods of the *Posterior Analytics* of which he was aware was that to be found in the brief section on demonstration in the dialectic handbook. See p. 48.

[2] See also [IV, 25–6].

[3] That is to say, in the syllogism, 'All men are mortal, Socrates is a man, therefore Socrates is mortal', not merely the major premise, the universal proposition, 'All men are mortal', is assumed, but equally the minor premise, 'Socrates is a man'. As well as assuming the general statement on which the conclusion must rest, the syllogiser must assume that his particular subject has been correctly classified ('Socrates' belongs to the class 'man'). Neither assumption can be verified by means of syllogistic. In the *Posterior Analytics* Aristotle noted that in demonstration the universal premise must be true of necessity, and correct application of the middle term known in advance, to ensure the truth of the minor premise.

being (as the physicians say) in the first digestion; which is not to be rectified by the subsequent functions. [IV, 411][1]

Bacon regarded even the obviously fallacious induction by simple enumeration as preferable to syllogistic for establishing the premises of scientific syllogisms. According to Bacon there is no problem in giving simple descriptions of observations, and hence induction, which generalises such descriptions to regularities does not involve a separate, fallible judgment procedure such as is invoked in the syllogism:

With regard however to judgment by induction there is nothing to detain us; for here the same action of the mind which discovers the thing in question judges it; and the operation is not performed by the help of any middle term, but directly, almost in the same manner as by the sense. For the sense in its primary objects at once apprehends the appearance of the object, and consents to the truth thereof. [IV, 428]

The comparison made in this passage is instructive. For Bacon, as I have previously noted, there appears to be no problem in the relation between descriptions of observations and the observations themselves, where the objects are 'primary' or fundamental enough. Between such objects there is a large enough degree of real resemblance to justify the immediate application of the universal term covering all such cases. In the same way Bacon believed that in any induction the simple description of an instance is unproblematic as long as it is given in terms which are primary, and no additional assumptions are needed to enable the investigator to reach his conclusion. Although he goes on to dismiss 'puerile' induction (induction by simple enumeration) as inadequate for establishing scientific truths on other counts [IV, 25, 70, 410, 428; III, 246], Bacon stresses from the outset that an induction of some sort is the solution of the problem:

In establishing axioms, another form of induction must be devised than has hitherto been employed; and it must be used for proving and discovering not first principles (as they are called) only, but also the lesser axioms, and the middle, and indeed all. For the induction which proceeds by simple enumeration is childish; its conclusions are precarious, and exposed to peril from a contradictory instance; and it

[1] See also [IV, 24, 49].

generally decides on too small a number of facts, and on those only which are at hand. [IV, 97]

Bacon does not, of course, suggest that the syllogism should be abandoned altogether in scientific reasoning.[1] It is the inadequacy of syllogistic for discovering first principles which causes him concern. In a letter to Redemptus Baranzano, in answer to some query (now lost to us) about the scope of the syllogism, Bacon puts his position as follows:

I do not propose to give up syllogism altogether. Syllogism is incompetent for the principal things rather than useless for the generality.

In the Mathematics there is no reason why it should not be employed.[2] It is the flux of matter and the inconstancy of the physical body which requires Induction; that thereby it may be fixed, as it were, and allow the formation of notions well defined...

In Physics, you wisely note, and therein I agree with you, that after the Notions of the first class and the Axioms concerning them have been by Induction well made out and defined, Syllogism may be applied safely; only it must be restrained from leaping at once to the most general notions; and progress must be made through a fit succession of steps. [XIV, 377][3]

When first principles are established, either by induction, or because they are 'placets' guaranteed true by divine inspiration, or are conventionally correct as part of some consistent set of rules (like the rules of chess), syllogistic may safely be used to derive consequences consistent with those principles:

[1] It was quite standard in the period to identify 'reasoning' with 'syllogising' (although there was controversy over whether all mathematical proofs could be reduced to syllogistic form). Bacon takes this for granted when he investigates the question 'do brutes possess the power of reason?' in the form 'do brutes have the power of syllogising?' [IV, 179]. Ramus identifies 'ratiocinatio' with syllogising in the *Dialecticae Institutiones* (Parisiis, 1543) thus: 'Syllogismus igitur...est argumenti cum quaestione firma, necessariaque collocatio, unde quaestio ipsa concluditur, atque aestimatur: latine ratiocinatio a Cicerone nominatur. συλλογίζομαι enim, et ratiocinor, eandem rem significant' (fo. 20r). 'Syllogism therefore...is the firm and necessary linking of 'arguments' with a 'quaestio', whence the 'quaestio' itself is concluded and assessed: in Latin this is called 'ratiocinatio' by Cicero, for 'I syllogise' and 'I reason' signify the same thing.'

[2] There was extensive controversy in the period on the question of the possibility of reducing mathematical proofs to syllogistic form. See H. Schüling, *Die Geschichte der axiomatischen Methode im 16. und beginnenden 17. Jahrhundert* (Hildesheim, 1969), chapter 9. Bacon apparently assumes that the first principles of mathematics are self-evidently true, that is, known with certainty.

[3] Whitaker notes that Bacon does not entirely reject syllogistic, and cites the Baranzano letter. See V. K. Whitaker, *Bacon and the Renaissance Encyclopedists* (unpublished Ph.D. diss., Stanford, 1933), pp. 70–1.

For after the articles and principles of religion have been set in their true place, so as to be completely exempted from the examination of reason, it is then permitted us to derive and deduce inferences from them according to their analogy [consistent with them]...Nor yet does this hold in religion alone, but also in other sciences both of a greater and smaller nature; namely, wherein the primary propositions are arbitrary and not positive [Placita sint, non Posita]; for in these also there can be no use of absolute reason. For we see in games, as chess or the like, that the first rules and laws are merely positive, and at will; and that they must be received as they are, and not disputed; but how to play a skilful and winning game is scientific and rational. [V, 114]

Indeed, Bacon is quite clear that within the terms of dialectic as an art of presentation and argument based on received or accepted premises, syllogistic is a perfectly adequate and appropriate tool:

For in the syllogism propositions are reduced to principles through intermediate propositions. Now this form of invention or of probation [support] may be used in popular sciences, such as ethics, politics, laws, and the like; yea, and in divinity also, because it has pleased God of his goodness to accommodate himself to the capacity of man; but in Physics, where the point is not to master an adversary in argument, but to command nature in operation, truth slips wholly out of our hands. [IV, 411][1]

The inductive method by means of which Bacon proposes to 'establish progressive stages of certainty' [IV, 40] in the sciences is designed to exploit the inherent capacities of the human faculties, and to bypass their errors or Idols. This approach to the problem of finding a method for arriving at sound first principles is consistent with Aristotle's remarks in *Posterior Analytics* II, 19, where he gives a brief stage by stage account of the process of cognition, and links this with scientific demonstration. Bacon outlines his procedure as follows:

Now my directions for the interpretation of nature embrace two generic divisions; the one how to educe and form axioms from experience; the other how to deduce and derive new experiments from axioms. The former again is divided into three ministrations; a ministration [aid] to the sense, a ministration to the memory, and a ministration to the mind or reason.

[1] See also [IV, 17, 24, 52, 112]. On the restricted aims of dialectic Bacon here refers to (the dialectic he was taught at Cambridge) see chapter 1, especially p. 30.

For first of all we must prepare a *Natural and Experimental History*, sufficient and good; and this is the foundation of all; for we are not to imagine or suppose, but to discover, what nature does or may be made to do.

But natural and experimental history is so various and diffuse, that it confounds and distracts the understanding, unless it be ranged and presented to view in a suitable order. We must therefore form *Tables and Arrangements of Instances*, in such a method and order that the understanding may be able to deal with them.

And even when this is done, still the understanding, if left to itself and its own spontaneous movements, is incompetent and unfit to form axioms, unless it be directed and guarded. Therefore in the third place we must use *Induction*, true and legitimate induction, which is the very key of interpretation. [IV, 127][1]

Bacon nowhere gives a unified account of the faculty psychology which he assumes as the basis for the proposed stages of the inductive method.[2] However, from scattered remarks in his writings it is possible to build up a fairly clear picture of a faculty psychology[3] which is thoroughly conventional in all but its emphasis on the predictability of human behaviour.

All knowledge of natural phenomena is gained via the senses. Smell, taste and touch respond to stimuli which impinge directly upon them (taste is a compound of touch and smell) [IV, 163–4]. Sight and hearing respond to distant stimuli, apparently without the communication of the substance of the source of impression [II, 430, 651; IV, 164, 356 (motion of impression)].

[1] For a particularly clear account of the relation between the various stages of the interpretation of nature and the faculties of the mind see [III, 553–6].

[2] As for instance does Hemmingius at the beginning of his treatise *De Lege Naturae* in which he uses faculty psychology as the basis for a theory of axiomatic ethics. See N. Hemmingius, *De Lege Naturae* (1577).

R. Hooke's scheme for remedying the defects of natural philosophy is closely modelled on Bacon's Great Instauration (Hooke acknowledges Bacon briefly p. 6). Hooke places the following as the first and second stages of 'the True Method of Building a Solid Philosophy, or of a Philosophical Algebra'; '1st: An Examination of the Constitution and Powers of the Soul, or an Attempt of Disclosing the Soul to its self, being an Endeavour of Discovering the Perfections and Imperfections of Humane Nature, and finding out ways and means for the attaining of the one, and of helping the other. 2dly, A Method of making use of, or employing these Means and Assistances of Humane Nature for collecting the Phenomena of Nature, and for compiling of a Philosophical History' (p. 7). See R. Hooke, 'A General Scheme, or Idea of the Present State of Natural Philosophy, and how its defects may be remedied by a Methodical Proceeding...', *Posthumous Works*, ed. Waller (London, 1705).

[3] On renaissance faculty psychology see G. S. Brett, *A History of Psychology* (London, 1921), vol. II ('Mediaeval and early modern period'); E. R. Harvey, *The Inward Wits: An Enquiry into the Aristotelian Tradition of Faculty Psychology in its literary relations during the later middle ages and the Renaissance* (unpublished Ph.D. diss., London, 1970).

Sense impressions are transmitted to the mind by animal spirits, which pervade the entire body [IV, 165; V, 268, 324, 413]. These animal spirits are the one line of communication between mind and body. They are the vehicle for transmission of all initial stimuli to the mind, and they translate the voluntary decisions of the mind into voluntary motions of the body [IV, 401; V, 358].

The animal spirits are also responsible for all *feelings*. Stimuli which are intrinsically pleasant, terrifying, or painful to human nature produce specific motions of the spirits which result both in characteristic postures of the body (gooseflesh and shaking for fear, for instance) and a mental response appropriate to the stimulus [II, 567ff.]. Bacon believed that the most direct effects on the animal spirits result from the stimulus of other spirits: drunkenness is accounted for as a perturbation of the animal spirits caused by the intrusion of the spirits of the wine, which disrupt the natural motions of the animal spirits, causing both lack of control over movement and hallucination [II, 571]. Bacon also believed that sounds produce a more direct effect on the body than smells, because the stimulus is more rarified and spirit-like [II, 389].[1] From Bacon's account of 'visibles and audibles' it appears that he imagined that every sensory stimulus is accompanied by a feeling of pleasure or pain in some degree, brought about by the motion of the spirits [II, 429, 630].

All bodies, whether animate or inanimate, are subject to the effects of external stimuli from other bodies [IV, 402]. Only animate bodies are *sensible* to such impressions, because of the mobility of the spirits which pervade them [V, 323; II, 528, 530]. All animate bodies react to stimulus as a result of movement of the animal spirits. In all animate bodies except man the reaction is direct; the animal spirits constitute the 'sensible soul',[2] and the 'inflammation' of these 'supplies peculiar motions and faculties' [V, 323], and prompts action. Man, however, possesses as an additional component in his response to external stimuli a mind or *rational soul*, to which the spirits (or sensible soul) are subservient [IV, 396–8].[3]

[1] For a similar view see D. P. Walker's account of Ficino's theory of spirits, Walker, *Spiritual and Demonic Magic from Ficino to Campanella* (London, 1958), pp. 3–11.

[2] Compare [IV, 398], where the sensible soul is described, with [V, 323], where the vital spirits are described.

[3] Bacon's theory of animal spirits markedly resembles that of Telesius. In fact, Bacon's faculty psychology as a whole is strongly reminiscent of Telesius' *De Rerum Natura*. For a

The rational soul was not brought into being at the creation when God organised matter into the natural order we know, but was separately created [IV, 396–7; VII, 221]. It is not, therefore, subject to the laws of nature [IV, 398; V, 314]. It possesses the capacity for complete knowledge of nature [III, 220, 265] (which paradoxically excludes knowledge of its own essential nature otherwise than by divine revelation [IV, 398]). The faculties of the rational soul are *memory*, *imagination* and *reason*. All these handle the information transmitted to them 'ex analogia hominis' rather than 'ex analogia universi', if left to their own devices, because the animal spirits substantially alter the message in the course of transmission [IV, 58–9, 325].

Memory is a simple store. Basically it contains the images of all external stimuli. Since these are singular, the contents of memory consist in the first place of singular instances, not generalisations:

> The sense, which is the door of the intellect, is affected by individuals only. The images of those individuals – that is, the impressions which they make on the sense – fix themselves in the memory, and pass into it in the first instance entire as it were, just as they come. [IV, 293]

These images stored in the memory provide the raw material for all mental operations.[1] It is, according to Bacon, the only reliable basis for any analysis of natural processes. Any higher level of generalisation which has not been achieved by some strictly regulated procedure (such as the inductive method) will inevitably have lost the primitive certainty of these basic images.

The memory does, of course, also store material learned from

summary of Telesian psychology see van Deusen, op. cit., pp. 53–71. On Bacon's theory of spirits and its relation to that of Telesius see D. P. Walker, 'Francis Bacon and *spiritus*', *Science, Medicine and Society in the Renaissance*, ed. Debus (New York, 1972), vol. II, pp. 121–130.

[1] Unlike most traditional faculty psychologies, Bacon's faculty psychology does not include a 'common sense' which amalgamates the responses of the individual sense organs to form one composite image for storing (there is a single reference to the 'common sense' [II, 423]). He may have imagined that each sense organ merely conveys a single aspect of the composite whole which is 'the event observed', and about whose description and unity there is no difficulty. Here as elsewhere Bacon's robust realism allows him to evade all philosophical problems of perception. Descriptions of all observed states and processes are lumped together as *instances*, and Bacon sees no problem about the relations between what occurs, what is perceived, and their descriptions. His failure to give any account of the way in which perception is built up of the separate responses of the sense organs is typical of the period. See e.g. Hemmingius, *De Lege Naturae*.

books, or from conversations with other men, the outcomes of previous reasoning processes, and the products of imaginative invention. These can be used as the basis for communication and disputation, but are ignored by Bacon in his discussion of scientific investigation, in favour of material suited to the search for certain knowledge. 'Individuals, which are circumscribed by place and time' [IV, 292] are the basic material for any intellectual process for deriving universal definitions of natural kinds, and the rules relating them.[1]

The animal spirits are the intermediary between external stimulus and mental response, and between intellectual decision to act and subsequent action. Imagination is the intermediary both between the earliest intellectual images and reason, and between rational assessment of a situation and the subsequent (ethical and emotional) decision to act:[2]

> The imagination performs the office of an agent or messenger or proctor in both provinces, both the judicial and the ministerial. For sense sends all kinds of images over to imagination for reason to judge of; and reason again when it has made its judgment and selection, sends them over to imagination before the decree be put in execution. For voluntary motion is ever preceded and incited by imagination; so that imagination is as a common instrument to both, – both reason and will.
>
> [IV, 405–6]

Imagination depicts in the mind images of individual past experiences, whose record has been stored in the memory. It enables images to be compared for similarity and dissimilarity, and hence is necessary for the generalising function of reason. Past experiences may also be combined in unconventional ways in the imagination, so as to create compound images which could never occur naturally (the Chimera, an animal compounded of various parts of naturally occurring animals is the traditional example of such 'feigning' on the part of imagination). Imagination always works with what has actually been experienced or described in the past. It cannot create the *components* of a fictitious compound image.

Once a decision has been rationally arrived at, an image of the

[1] For Bacon's views on mnemonic methods see p. 71.

[2] By the sixteenth century, *imaginatio* had become generally accepted as the single, all purpose faculty mediating between sense and reason, replacing the complex mediaeval systems of internal faculties. See K. Park, *The Imagination in Renaissance Psychology* (M. Phil. diss., The Warburg Institute, University of London).

desired outcome of action must be generated in the imagination before any voluntary motion takes place:

> The imagination is as it were the director and driver of [voluntary] motion, insomuch that when the image which is the object of the motion is withdrawn the motion itself is immediately interrupted and stopped (as in walking, if you begin to think eagerly and fixedly of something else, you immediately stand still). [IV, 401]

In a similar way, the image of some situation to which a particular response is appropriate will produce the sensations and beginnings of action appropriate to it, even in the absence of the real event:

> Those effects which are wrought by the percussion of the sense, and by things in fact, are produced likewise in some degree by the imagination. Therefore if a man see another eat sour or acid things which set the teeth on edge, this object tainteth the imagination; so that he that seeth the thing done by another, hath his own teeth also set on edge. So if a man see another turn swiftly and long, or if he look upon wheels that turn, himself waxeth turn-sick. So if a man be upon an high place without rails or good hold, except that he be used to it, he is ready to fall: for imagining a fall, it putteth his spirits into the very action of a fall. [II, 598]

Because it is a natural characteristic of the imagination to fashion systems 'according to the pleasure of the mind' [V, 504] the imagination presents a potential obstacle to understanding of nature. For instance:

> The human understanding is moved by those things most which strike and enter the mind simultaneously and suddenly, and so fill the imagination; and then it feigns and supposes all other things to be somehow, though it cannot see how, similar to those few things by which it is surrounded. [IV, 56]

One of the undertakings of the interpretation of nature is to protect the investigator against the misleading systems which he is inclined to build in his imagination on the strength of half-formed theories about the world.

By comparing and combining images of singular past experiences imagination creates fantastic and unnatural composite events and objects, and systems of order 'ex analogia hominis'.

Reason compares and combines the same past impressions, but it does so according to the rules of nature ('ex analogia universi'), and not according to the whim of the individual:

For the images of individuals are received by the sense and fixed in the memory. They pass into the memory whole, just as they present themselves. Then the mind recalls and reviews them, and (which is its proper office) compounds and divides the parts of which they consist. For the several individuals have something in common one with another, and again something different and manifold. Now this composition and division is either according to the pleasure of the mind, or according to the nature of things as it exists in fact. If it be according to the pleasure of the mind, and these parts are arbitrarily transposed into the likeness of some individual, it is the work of imagination; which, not being bound by any law and necessity of nature or matter, may join things which are never found together in nature and separate things which in nature are never found apart; being nevertheless confined therein to these primary parts of individuals. For of things that have been in no part objects of the sense, there can be no imagination, not even a dream. If on the other hand these same parts of individuals are compounded and divided according to the evidence of things, and as they really show themselves in nature, or at least appear to each man's comprehension to show themselves, this is the office of reason; and all business of this kind is assigned to reason. [V, 503–4]

A basic task of reason is to derive the generalisations which enable men to predict future events, and to make decisions about future actions, on the basis of individual past events. It does not, according to Bacon, handle images (pictures of events and processes), but *notions* [IV, 292]. That is to say, it is a *verbal* faculty, manipulating the terms which denote objects, rather than the images of those objects. Reason manipulates notions so as to discover the widest possible groupings of objects into natural kinds, and the universal rules governing natural processes [IV, 292–3].

Following rational assessment of a situation which necessitates action the image of its outcome prompts an act of *will*, which sets in motion the animal spirits to produce a voluntary action. Although Bacon uses the terms *appetite* and *will* consistently to refer to unconsidered and considered promptings to action, he does not assign to either a physical position in the mind or brain, as he apparently does to memory, imagination and reason [IV,

378]. Appetite and will are terms attached to particular activities of the mind, involving one or more of the faculties.[1]

If man's mind had not been corrupted at the Fall, an act of will would prompt man to action whenever reason judged that action to be to his overall good. Present and immediate pleasures, however, appear preferable to long-term and absolute good. Persuasion may often be necessary to make a man act for the best:

> If the affections themselves were brought to order, and pliant and obedient to reason, it is true there would be no great use of persuasions and insinuations to give access to the mind, but naked and simple propositions and proofs would be enough. But the affections do on the contrary make such secessions and raise such mutinies and seditions...that reason would become captive and servile, if eloquence of persuasions did not win the imagination from the affections' part, and contract a confederacy between the reason and imagination against them. For it must be observed that the affections themselves carry ever an appetite to apparent good, and have this in common with reason; but the difference is that affection beholds principally the good which is present; reason looks beyond and beholds likewise the future and sum of all. [IV, 456–7]

This holds equally in one's own actions, and in counselling others to act.[2]

As noted above, the animal spirits mediate between sense experience and the rational soul. The spirits respond to sense stimuli, and by their motion produce both affective states and mental images. This account provides an entirely stimulus/response explanation for the way in which the hiatus between body and rational soul is bridged. Although the body and soul are not subject to the same laws, the reaction of the rational soul is a direct consequence of the body's obeying natural laws.[3]

[1] Bacon uses the term '*facultas*' both to describe the *parts* of the mind, memory, imagination and reason, and the mental operations carried out by one or more of the faculties. Wallace has collected together Bacon's scattered pronouncements on faculty psychology, but treats appetite and will as faculties on a par with reason and imagination. See K. Wallace, *Francis Bacon on the Nature of Man* (Illinois, 1967).

[2] This is a view of the relation between the affections and reason traditionally used to justify the use of rhetoric in oratory, to persuade where reason cannot. See for instance T. Wilson's introduction to his *Arte of Rhetorique* (1553).

[3] Walker comments on the same feature of Telesius' theory of spirits and their mediation between body and mind: '[Telesius] is not using spirit as a bridge-concept, but in order to account for centralized systems of activity, particularly animals and men' (op. cit., p. 190). For both Bacon and Telesius 'everything is both sentient and extended' (Walker, p. 190), and the free movement of animal spirits in animals and man accounts for their coor-

Bacon explores the relation between bodily stimuli, animal spirits and mental response further under the heading of the *Doctrine of the League* [IV, 375ff.]. Bacon believed that there is a fixed and determinable relation between each bodily state and a corresponding state of mind and affections. A given bodily state indicates a specific state of mind; a given state of mind has particular bodily effects. It is therefore possible to diagnose accurately an individual's state of mind by scrutinising his posture and facial expression [IV, 376]. There is, therefore, a limit to the amount of dissembling of true emotions which can be achieved by external posturing.

By the same token, a given state of the body may induce a state of mind frequently associated with it. For example, dreams are induced by particular states of the sleeper's body:

> when the same sensation is produced in the sleeper by an internal cause which is usually the effect of some external act, that external act passes into the dream. A like oppression is produced in the stomach by the vapour of indigestion and by an external weight superimposed; and therefore persons who suffer from the nightmare dream of a weight lying on them, with a great array of circumstances. A like pendulous condition of the bowels is produced by the agitation of the waves at sea, and by wind collected round the diaphragm; therefore hypochondriacal persons often dream that they are sailing and tossing on the sea. [IV, 377]

Bacon also believed that it is possible to prescribe remedies for the body which can correct such disorders of the mind as melancholy, and that states of health of the body can be altered by encouraging particular trains of thought.

Bacon's view of the close and predictable relation between body and rational soul is important because it sets study of man alongside that of other of phenomena. Man's body and sensible soul (spirits) are susceptible to the same sort of analysis as all other natural bodies [V, 219]. Such an analysis will yield rules governing involuntary behaviour which are as rigid and incontrovertible as those governing the growth of plants. And because the animal spirits are directly linked with the rational soul, even rational, voluntary behaviour is constrained not to violate the

dinated actions in reponse to stimuli. In animistic theories like Telesius' and Bacon's the problem of how an immaterial soul can impart motion to material spirits does not arise, because the mechanistic hypothesis that all causation occurs through imparted motion is not clearly stated.

rules which govern the sensible soul. Although Bacon insists that the rational soul (and hence God's divine intervention) provides the governing force in man, this theory goes a long way towards usurping the prerogative of man's spiritual part over his earthly part.[1]

The stages of the inductive method, which supposedly correspond to the stages of human cognition, are the compilation of natural histories, the tabulation of their contents, and the carrying out of an induction by elimination between the contents of the tables. The first of these stages controls the initial acquisition of sense data; the natural histories ensure that only primitive, unprocessed observations received directly in the mind from the senses *via* the animal spirits are used as the basis for investigation. The second stage is the preliminary classification of this basic material into types; Bacon's tabulation procedure and use of topics of investigation controls the distorting influences of memory left to its own devices [III, 553]. The final stage is the rational procedure whereby this tentatively grouped material gives rise to true generalisations; the induction replaces syllogistic for this purpose, and 'rises by gradual steps to that which is prior and better known in the order of nature', instead of 'establishing certain abstract and useless generalities' [IV, 50].

Bacon's classification of knowledge

In his classification of knowledge Bacon again uses the faculties of memory, imagination and reason to justify his division of subjects:[2]

[1] Bacon covers himself against any charge of heterodoxy with a sophistical remark at the end of his discussion of the doctrine of the league: 'But if any man of weak judgment conceive that these impressions of the body on the mind either question the immortality of the soul, or derogate from its sovereignty over the body,...Let him take the case...of monarchs who, though powerful, are sometimes controlled by their servants, and yet without abatement of their majesty royal' [IV, 377–8]. Telesius and his pupil Donius, who are Bacon's avowed influences in the 'animal spirits' aspects of his faculty psychology [IV, 398, 325] both introduce an immaterial, transcendent soul at a late stage, and their treatment of spirits laid them open to attack by the church. See Walker, op. cit., pp. 189–94.

[2] On renaissance innovations in the classification of knowledge see R. Flint, *Philosophy as Scientia Scientiarum, and A History of Classifications of the Sciences* (Edinburgh and London, 1904); P. O. Kristeller, 'The modern system of the Arts', *Renaissance Thought II* (New York, 1965), pp. 163–227, and 'Renaissance philosophy and the medieval tradition', *Renaissance Concepts of Man* (New York, 1972), pp. 110–55. According to Flint Campanella bases his classification of knowledge on *history*, the record of immediate perceptions (p. 101). See Campanella, *Universalis Philosophiae* (Parisiis, 1638), Book V. On Campanella's classification of knowledge see L. Blanchet, *Campanella* (Paris, 1920), p. 231ff.

The best division of human learning is that derived from the three faculties of the rational soul, which is the seat of learning. History has reference to the Memory, poesy to the Imagination, and philosophy to the Reason...That these things are so, may be easily seen by observing the commencements of the intellectual process. The sense, which is the door of the intellect, is affected by individuals only. The images of those individuals – that is, the impressions which they make on the sense – fix themselves in the memory, and pass into it in the first instance entire as it were, just as they come. These the human mind proceeds to review and ruminate; and thereupon either simply rehearses them, or makes fanciful imitations of them, or analyses and classifies them. Wherefore from these three fountains, Memory, Imagination, and Reason, flow these three emanations, History, Poesy, and Philosophy; and there can be no others. For I consider history and experience to be the same thing, as also philosophy and the sciences. [IV, 292–3]

Although the material of theology is directly inspired into the mind by God, and does not pass through the normal processes of cognition, it also falls into the classification according to the faculties:

Theology therefore in like manner consists either of Sacred History, or of Parables, which are a divine poesy, or of Doctrines and Precepts, which are a perennial philosophy. [IV, 293]

It does not do to analyse too closely the detailed fit of the academic subjects which Bacon discusses and the place in the classification to which he assigns them. For instance, although natural history fits the specifications for the subject corresponding to the basic operation of memory, Bacon's treatment of civil history shows that within this subject generalisation and interpretation are allowed to play a part (see below, chapter 8). The classification does, however, provide a particularly good setting for *philosophy*. Philosophy covers all that body of human knowledge which is the work of reason:

For under philosophy I include all arts and sciences, and in a word, whatever has been from the occurrence of individual objects collected and digested by the mind into general notions. [V, 504]

The most important consequence of this particular way of classifying is that it avoids the initial division, basic to mediaeval and early-sixteenth-century classification, of knowledge into theoretical and practical.[1] In Bacon's classification the theoretical and practical aspects of each branch of philosophy are bracketed together. This enables Bacon to make the point firmly at the outset of his survey of the existing state of knowledge, that in each field the theoretical study and its practical consequences ought to be closely associated, instead of being placed at opposite poles with the practical studies markedly inferior to the theoretical. For Bacon all human knowledge, that is, all the contents of philosophy, is to be judged by its effectiveness in action [IV, 110]. The certain first principles which are the goal of Baconian science are to be at once contemplative and operative; universally true definitions and rules for operating on natural phenomena:

[1] Aristotle divided knowledge into speculative, practical and productive. See *Nicomachean Ethics* VI, 3–7 (1139b14-1141b25); also *Topics* VI, 6 (145a16); VIII, 1 (157a10); *Metaphysics* I, 1 (981b25-982a1); VI, 1 (1025b19-27); IX, 2 (1045b3); XI, 7 (1064a10-19); *Eudemian Ethics* (1216b11-17). Speculative knowledge is knowledge of things in themselves; practical knowledge is knowledge of things as they are conducive to man's ethical good; and productive knowledge is knowledge of things useful to man.

On Aristotelian classifications of knowledge in the middle ages see L. Baur, 'Dominicus Gundissalinus *De Divisione Philosophiae*: herausgegeben und philosophiegeschichtlich untersuch nebst einer Geschichte der philosophischen Einleitung bis zum Ende der Scholastik', *Beiträge zur Geschichte der Philosophie des Mittelalters*, ed. Baeumker and von Hertling, Bd. IV, Heft 2–3 (Münster, 1903). Baur gives a comprehensive survey of scholastic classifications of knowledge and their classical sources.

The primary division was into theoretical and practical knowledge. Theoretical knowledge is divided into Natural Philosophy (the study of bodies, and the qualities of matter), Metaphysics or *Philosophia Prima* (the study of what can be considered apart from matter) and Mathematics (the study of quantities abstracted from matter). Abstract and concrete physics, study of what Bacon calls latent configuration and process (see p. 141), and various types of motion, fall under Natural Philosophy; study of principles and terms common to all branches of knowledge, intelligences, God, etc., fall under Metaphysics; study of arithmetic, geometry, music and astronomy, fall under Mathematics. The divisions of practical knowledge are Ethics, Economics and Politics. Theoretical knowledge is the province of *sagacitas*, practical of *prudentia*.

The *logical arts* (grammar, rhetoric, poesy, topics or dialectic) are sometimes considered as part of philosophy, sometimes as instruments of philosophy, sometimes as a separate category – arts of discourse. The *productive* or *mechanical arts* (agriculture, navigation, etc.) are sometimes considered as a branch of Economics, sometimes as an independent study, sometimes associated with natural philosophy.

For examples of early-sixteenth-century classifications of knowledge which follow the mediaeval pattern see G. Savonarola, 'Opus de divisione, ordine, ac utilitate omnium scientiarum, in poeticen apologeticum', *Compendium totius philosophiae, tam naturalis quam moralis* (Venetiis, 1534); G. Reisch, *Margarita Philosophica* (Friborgi [i.B.], 1504). I reproduce Reisch's diagrammatical representation of his classification (fo. 3r) for comparison with G. Watts' representation of Bacon's classification from his 1640 edition of the *De Augmentis*.

See also G. M. Paré, A. Brunet and P. Tremblay, *La renaissance du XIIe siècle* (Paris, 1933), chapter III, 'Matières et Procédés d'Enseignement', pp. 94–137; and J. A. Weisheipl, 'Classification of the sciences in mediaeval thought', *Mediaeval Studies*, 27 (1965), pp. 54–90.

Human knowledge and human power meet in one; for where the cause is not known the effect cannot be produced. Nature to be commanded must be obeyed; and that which in contemplation is as the cause is in operation as the rule. [IV, 47]

(See also [IV, 146; III, 553–4].) Operative and contemplative knowledge are faces on one coin, bound together under philosophy as alternative aspects of a rational understanding of nature.[1]

In this context Bacon takes over the scholastic terminology of 'ascending' and 'descending' methods (see chapter 1, p. 34) to represent the relationship between the detailed intellectual procedure for arriving at first principles (ascending to what is better known in nature), and that for deriving operative power from those same principles (descending to what is better known to us); intellectual procedures whose practical details must, according to Bacon, differ in some respects:

And certainly though I may seem to say this in sport, yet I think a division of this kind most useful, when propounded in familiar and scholastical terms; namely, that the doctrine of Natural Philosophy be divided into the Inquisition of Causes, and the Production of Effects; Speculative and Operative. The one searching into the bowels of nature, the other shaping nature as on an anvil. And though I am well aware how close is the intercourse between causes and effects, so that the explanations of them must in a certain way be united and conjoined; yet because all true and fruitful Natural Philosophy has a double scale or ladder, ascendent and descendent, ascending from experiments to axioms, and descending from axioms to the invention of new experiments; therefore I judge it most requisite that these two parts, the Speculative and the Operative, be considered separately, both in the intention of the writer and in the body of the treatise. [IV, 343]

For our road does not lie on a level, but ascends and descends; first ascending to axioms, then descending to works. [IV, 96]

But my course and method, as I have often clearly stated and would wish to state again, is this, – not to extract works from works or experiments from experiments (as an empiric), but from works and experiments to extract causes and axioms, and again from those causes and axioms new works and experiments, as a legitimate interpreter of nature. [IV, 104][2]

[1] Although the two branches are extremely closely associated Bacon does indicate that different skills are required for the contemplative and operative branches of natural philosophy. Operative knowledge requires 'prudentia'; contemplative knowledge requires 'sapientia' [III, 554].

[2] Crombie and Kosman have pointed out the parallels between Bacon's 'ascending and descending ladder of axioms' and the traditional Aristotelian treatment of composition

Whereas both the 'ascent' and 'descent' of traditional demonstration are part of the *theoretical* study of natural philosophy, Bacon's 'ascent' links works (practical knowledge) with axioms (theoretical knowledge) [IV, 50], and his 'descent' links axioms with further works [III, 556].

Metaphysic is for Bacon the supreme theoretical study of natural phenomena; the study concerned with universal and unchanging first principles. But because of his insistence on the identity of definitions of first principles and productive instructions – the contemplative and operative faces of definitions – Bacon envisages a close link between *Metaphysic* and *Physic*, traditionally a separate study of bodies, rather than of properties abstracted from bodies. Abstract physics, for Bacon, provides local generalisations and rules for transforming bodies under some circumstances (following natural processes) whereas the definitions and rules of metaphysic are universal and independent of circumstances:

> Abstract Physics may most rightly be divided into two parts – the doctrine concerning the Configurations of Matter, and the doctrine concerning Appetites and Motions...[T]o inquire the form of dense, rare, hot, cold, heavy, light, tangible, pneumatic, volatile, fixed, and the like, as well configurations as motions, which in treating of Physic I have in great part enumerated (I call them *Forms of the First Class*) ...this, I say, it is which I am attempting, and which constitutes and defines that part of Metaphysic of which we are now inquiring. Not but that Physic takes consideration of the same natures likewise (as has been said); but that is only as to their variable causes. [IV, 355, 361]

Physics moves from works to further works *via* the intermediary of local generalisations; metaphysic moves from works to axioms and on to works of further power. Bacon thus envisages a pyramidal structure for natural philosophy, with natural history at the base, metaphysic at the apex, and physics in between.

Bacon's classification groups together as the outcomes of the reasoning process natural theology, natural philosophy, politics and ethics, as branches of philosophy. All these, Bacon is at pains to stress, are branches of natural knowledge whose fields

and resolution. Kosman has undertaken a detailed study of mediaeval and renaissance theories of natural priority and has interpreted Bacon's inductive method in the light of this background. Both authors appear to oversimplify the issues in their treatment of renaissance discussions of method. See A. C. Crombie, *Robert Grosseteste and the origins of experimental science 1100–1700* (Oxford, 1953), pp. 301–3; Kosman, op. cit., pp. 257–312.

overlap and interact, and which share fundamental tenets which underlie all natural processes:

> And generally let this be a rule; that all divisions of knowledges be accepted and used rather for lines to mark and distinguish, than sections to divide and separate them; in order that solution of continuity in sciences may always be avoided. For the contrary hereof has made particular sciences to become barren, shallow, and erroneous; not being nourished and maintained and kept right by the common fountain and aliment. [IV, 373]

The sciences must not simply constitute internally consistent systems; each must confirm the theories of adjacent and related sciences. All arts (the applied counterparts of the individual sciences) must be closely linked with the whole of natural philosophy [IV, 79].

It will be noted, by comparing Bacon's classification as schematised by Watts (1640) with Reisch's classification from the *Margarita Philosophica* (see pages 102 and 103) that the studies of ethics and politics acquire in Bacon's scheme a theoretical branch in addition to its traditional practical branch. In this way 'human philosophy', the study of man, is brought into line with natural philosophy, and is separated from inspired theology.

Philosophia prima

Bacon believed that there are some general principles which recur and stand out in so many separate sciences that they may be immediately judged to be universally true. This is a standard Aristotelian view.[1] The indemonstrable propositions which are the basis for any art or science fall into two classes. Some principles are *propria*, and belong to a single science (for example, the principle of geometry, 'a line is a longitude without latitude'). Others are *communia*, common to all sciences (for example, the principle, 'take equals from equals and equals remain').[2] *Principia communia* of their nature tend to be general relational propositions rather than definitions, because on the whole it is propositions about relations between components which are independent of any particular material. Bacon collects

[1] One of the sources is *Posterior Analytics* I, 10 (76a30-77a5).

[2] I have taken these examples from Caesarius' account 'De principiis et eorum multiplicia differentia', in tractate VII, 'De Demonstratione'; *Dialectica* (ed. cit.), fo. T2v-T3v. See also Melanchthon, *Erotemata Dialectices* (Witebergae, 1562 ed.), pp. 239–40.

The Emanation of SCIENCES, from the Intellectual Faculties of MEMORY, IMAGINATION, REASON.

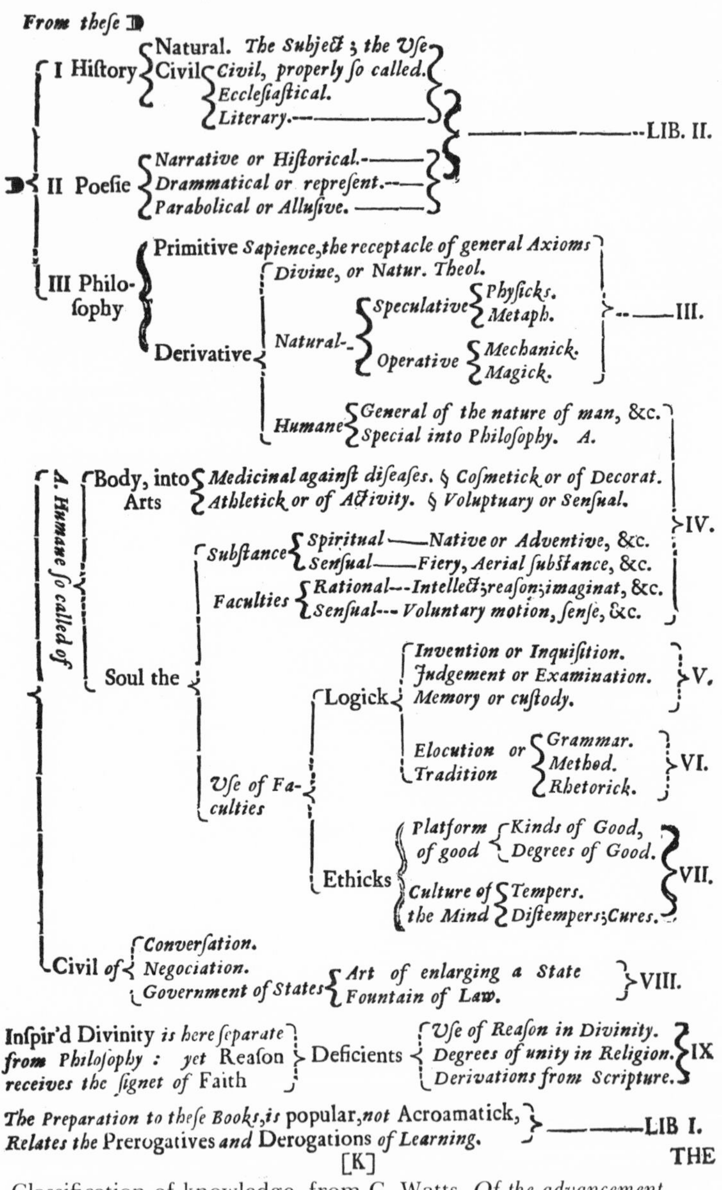

Classification of knowledge, from G. Watts, *Of the advancement and proficience of learning* (Oxford, 1640).

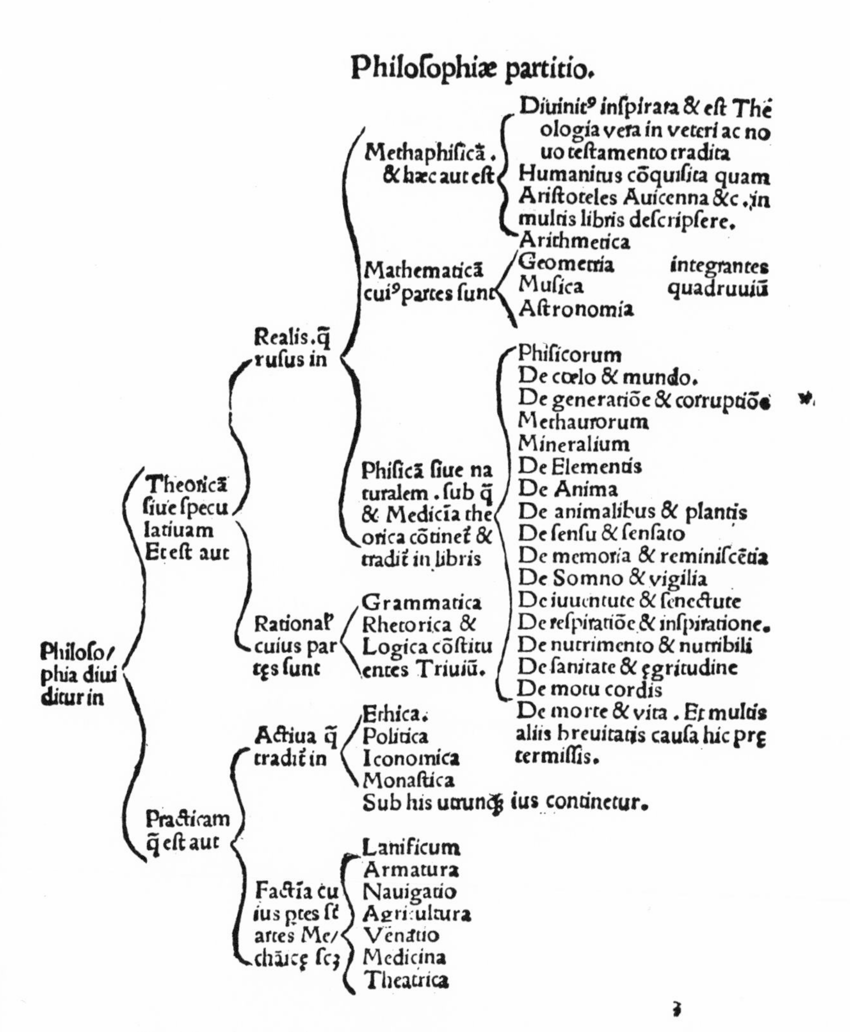

Classification of knowledge, from G. Reisch, *Margarita Philosophica* (Friborgi [i.B.], 1504).

such principles together in a separate category from his *forms* (essential definitions), which he calls *philosophia prima*.[1] However, Bacon is more than conventionally bold in his identification of principles common to all sciences. He is also particularly anxious to show how the three branches of philosophy share a basis of extremely general principles, thus reducing the amount of independent investigation needed in any one field.

Together with the most traditional examples, like, 'if equals be added to unequals the whole will be unequal' [IV, 337], Bacon gives such examples of *philosophia prima* as the following:

> 'The nature of everything is best seen in its smallest portions,' is a rule in Physics of such force that it produced the atoms of Democritus; and yet Aristotle made good use of it in his Politics, where he commences his inquiry of the nature of a commonwealth with a family...'Things are preserved from destruction by bringing them back to their first principles,' is a rule in Physics; the same holds good in Politics (as Macchiavelli rightly observed), for there is scarcely anything which preserves states from destruction more than the reformation and reduction of them to their ancient manners...'A discord ending immediately in a concord sets off the harmony,' is a rule in Music. The same holds in Ethics and in the affections... [IV, 338–9]

As these examples show, Bacon is at pains to show that principles of the 'exact' sciences (physics, music, mathematics) have their counterparts in the 'inexact' sciences (ethics and politics). This both supports his view that sciences can be systematised, and facilitates this systematisation, since direct investigation in ethics and civics is hampered by the fact that these subjects are 'immersed in matter, and with most difficulty reduced to axioms' [V, 32]. *Philosophia prima* will allow principles to be transferred from physics to ethics (say), in the certainty that what holds in the one will hold in the other.[2]

[1] In scholastic classifications of knowledge *philosophia prima* is a synonym for metaphysics. Bacon explicitly prefers the term *philosophia prima* for the study of the common axioms of the sciences and of the properties and relations which recur in all branches of knowledge, and transfers the term *metaphysic* to the study of forms (abstract or universal physics in the scholastic classifications): 'It appears by that which has been already said, that I intend Primitive or Summary Philosophy [Philosophia Prima] and Metaphysic, which heretofore have been confounded as one, to be two distinct things. For the one I have made a parent or common ancestor to all knowledge; the other, a branch or portion of Natural Philosophy' [IV, 345–6].

[2] In the *Erotemata Dialectices* Melanchthon endorses the transferring of *principia* from physics to ethics in a limited way as follows: 'Principiorum alia aliis illustriora, aut etiam priora sunt, Geometrae prima et maxime illustria vocant κοινὰς ἐννοίας, καὶ ἀξιώματα,

In the *Valerius Terminus* [III, 229] Bacon calls the principles of *philosophia prima*, '*Maxims*', because their relation to principles of a particular science is analogous to the relation which legal maxims bear to particular cases in law. Despite the local and specific character of individual judgments, general principles of equity can be seen to run through the legal code as a whole:

> For sciences distinguished have a dependence upon universal knowledge to be augmented and rectified by the superior light thereof, as well as the parts and members of a science have upon the *Maxims* of the same science, and the mutual light and consent which one part receiveth of another. [III, 229][1]

A significant aspect of Bacon's view of these principles or maxims is that it leads him to consider what to us appear verbal similarities as suggestive of common principles applying in different branches of knowledge. The most striking example amongst the general principles which he cites in the *De Augmentis* is that of *corruption*. The general principle is: 'Putrefaction is more contagious before than after maturity'. In physics, Bacon claims, this principle is apparent from observation. He also claims it as a principle of ethics in the following form:

posteriora vocantur hypotheses, alias aliter. Philosophia moralis ex Physica assumit hanc propositionem, Natura hominis condita est ad certum finem. Id in Physicis demonstratur, non confirmatur in Philosophia morali. Sed recipitur tanquam hypothesis, id est, firma propositio, aliunde assumpta, ut sit inchoatio doctrinae moralis' (ed. cit., p. 239).

A possible source for Melanchthon's and Bacon's views on the possibilities for transferring rules from one field to another is the tradition of biblical exegesis. The tradition of historical, allegorical, anagogical and tropological explanation of a single biblical passage may well have fostered the habit of looking at a single sentence for its significance in a variety of contexts. See B. Smalley, *The Study of the Bible in the Middle Ages* (Oxford, 1952), especially pp. 242–63; H. F. Dunbar, *Symbolism in Mediaeval Thought* (Yale, 1929), chapter 4.

In the *Advancement of Learning* (and *De Augmentis* Book I [I, 466]) Bacon ascribes one of his general principles of *philosophia prima* to 'the rabbins' ('Putrefaction is more contagious before maturity than after' [III, 298; IV, 338]), referring to the Jewish tradition of biblical exegesis. Philo interprets the Mosaic law about leprosy to which Bacon refers (Leviticus xiii, 12–13) in terms of the corruption of the soul. See *Quod Deus Immutabilis Sit, Works* (Loeb ed.) III, 73. In *The Zohar* this law is similarly interpreted, and transferred from physical to ethical context. See *The Zohar*, transl. Sperling and Simon (London, 1949) V, 9–33. See also below, chapter 11. I am indebted to J. Weinberg for these references.

[1] P. H. Kocher uses this analogy in support of the thesis that Bacon proposed an inductive method for use in jurisprudence. Kocher's argument unfortunately rests on a failure to discriminate between the fundamental definitions which are the goal of Baconian induction, and provisional generalisations in particular fields, which for the time being replace such certain definitions. See Kocher, 'Francis Bacon on the Science of Jurisprudence', *Journal of the History of Ideas*, 18 (1957), pp. 3–26, reprinted in Vickers ed., *Essential Articles for the study of Francis Bacon* (Connecticut, 1968), pp. 167–94. On generalisation, as opposed to induction, see below, chapters 7 and 8.

Men who are more wicked and profligate produce less corruption in the public manners than those who appear to have some soundness and virtue in them, and are only partly evil. [IV, 338]

In this case what we might regard as a figurative resemblance is considered by Bacon to mark an essential resemblance.[1] He insists that in many cases principles employing similar forms of words reveal an underlying general principle, rather than mere figurative resemblance:

Neither are all these things which I have mentioned, and others of this kind, only similitudes (as men of narrow observation may perhaps conceive them to be), but plainly the same footsteps of nature treading or printing upon different subjects and matters. [IV, 339]

Like its scholastic counterpart, Bacon's *philosophia prima* includes in addition to the study of these recurrent principles the investigation of certain pairs of general terms like *more, less*; *before, after*; *like, unlike*; *part, whole*; *equal, unequal*, which recur again and again in descriptions of natural kinds and processes.[2] Bacon calls such terms the *Adventitious Conditions of Essences*, or the *Common Adjuncts of Things*.[3] The first part of *philosophia prima* concerns the study of general principles which recur in all branches of philosophy; the second part concerns the study of general terms which recur in principles in all fields. Bacon holds that *philosophia prima* should investigate the application of these logical connectives (as he considers them to be) in the description of nature:

Let me call to mind then what I said above (in speaking of Primitive or Summary Philosophy [Philosophia Prima]) touching the Transcendental or Adventitious Conditions or Adjuncts of Essences. These are

[1] In *De Augmentis* I (*Advancement of Learning* I) this appears as follows: 'Take a view of the ceremonial law of Moses; you shall find, besides the prefiguration of Christ, the badge or difference of the people of God, the exercise and impression of obedience, and other divine uses and fruits thereof, that some of the most learned Rabbins have travelled profitably and profoundly to observe, some of them a natural, some of them a moral, sense or reduction of many of the ceremonies and ordinances. As in the law of leprosy, where it is said, *If whiteness have overspread the flesh, the patient may pass abroad for clean; but if there be any whole flesh remaining, he is to be shut up for unclean;* one of them noteth a principle of nature, that putrefaction is more contagious before maturity than after: and another noteth a position of moral philosophy, that men abandoned to vice do not so much corrupt manners, as those that are half good and half evil' [III, 297–8].

[2] Sources in Aristotle are *Metaphysics* IV, 2, 3 (1004a1-1005b35); IX, 3, 4. See also Plato, *Theaetetus* 185C-D.

[3] 'Adventitious' and 'common' here indicate that such properties occur in all kinds of body rather than being proper and essential to some particular kind.

Greater, Less, Much, Little, Before, After, Identity, Diversity, Potential, Actual, Habit, Privation, Whole, Parts, Active, Passive, Motion, Rest, Entity, Nonentity, and the like. And first let the different ways which I mentioned of viewing these things be remembered and noted; namely that they may be inquired either physically or logically. Now the physical handling of them I referred to Primitive or Summary Philosophy. There remains then the logical. [IV, 430]

(See also [IV, 339].) Correct application of such terms falls under logic; investigation of how such terms are used to describe nature falls under *philosophia prima*.[1]

Bacon includes the pair *motion, rest,* within his list of common adjuncts of things [IV, 430]. As he makes clear elsewhere, all tendencies of bodies, including an *inclination towards* some state differing from the present, are compounded of motions [IV, 356]. This standard Aristotelian use of the term *motion* for any change of state or disposition to change sustains the view that the same general principles can be manifested in the physical and ethical sciences. Bacon's justification for the principle, *Whatsoever is preservative of a greater Form is more powerful in action* [IV, 338], in its application to physical bodies and in politics, rests on the possibility of regarding as equivalent forms the following. In physics:

The Motion of Connexion, which regards communion with the universe, is stronger than the Motion of Gravity, which regards only communion with dense bodies. [IV, 232]

In politics:

That appetites which aim at a private good seldom prevail against appetites which aim at a more public good, except in small quantities. [IV, 232]

The study of universal principles enables the investigator tentatively to chart the relations between the sciences, to establish their general basis prior to the discovery of certain prin-

[1] The logical treatment of these types of terms in dialectic (including the striking pair, 'motion, rest') falls under the *post-predicaments.* See for instance the end of the third tractate of Peter of Spain's *Summulae*, or Seton, *Dialectica* (Londini, 1584), fo. E7v: 'Dicuntur post praedicamenta, quia summa genera eorum, ut opposita, prius, simul, idem et motus non possunt in praedicamentis collocari.' Or they are treated under *relatives* and *actio & passio,* amongst the *predicaments.* See Melanchthon, *Erotemata Dialectices* (Wittebergae, 1555), p. 72. Melanchthon gives the following 'relatives from quantity': 'Par Impar, Aequale Inaequale, Proportio, Totum Dimidium, Duplum Triplum &c.', and on the following page, under 'actio & passio' he lists the six species of motion: 'Generatio Corruptio, Augmentatio Diminutio, Alteratio Motus localis' (p. 74).

ciples by the inductive method. In the next chapter we shall see that 'forms', the essential definitions which are the goal of Bacon's inductive method, appear to be made up of terms which correspond to the adventitious conditions of essences. Hence, despite Bacon's claims that the inductive method rests on experience alone, it appears that the investigator is to be armed with knowledge of *philosophia prima* at the outset. *Philosophia prima*, the 'parent or common ancestor to all knowledge' [IV, 345], confirms the economy of natural order and is a prerequisite for the interpretation of natural phenomena:

> Since the divisions of knowledge are not like several lines that meet in one angle; but are rather like branches of a tree that meet in one stem (which stem grows for some distance entire and continuous, before it divide itself into arms and boughs); therefore it is necessary before we enter into the branches of the former division [of philosophy], to erect and constitute one universal science, to be as the mother of the rest, and to be regarded in the progress of knowledge as portion of the main and common way, before we come where the ways part and divide themselves. [IV, 337]

5
The goal of the interpretation of nature

In this chapter I give an account of those aspects of Bacon's natural philosophy which are necessary for an understanding of the inductive method. The account which follows is therefore highly selective, concentrating on his theory of forms (which are the goal of the inductive method) rather than on the details of his often idiosyncratic scientific beliefs.

Simple natures and their forms

Bacon starts out with the assumption that certain fundamental qualities of natural bodies can be seen by the shrewd investigator to be particularly widely distributed in nature. He assumes that such qualities can easily be identified, although occasionally it may be difficult to be sure about limiting cases. *Heat*, for instance, is a quality which occurs in many different kinds of body. The investigator is immediately aware of this, although until he knows what heat *is* (its essential definition) he may not always assign the term with complete reliability. Such widely distributed qualities Bacon calls *simple natures, cardinal virtues*, or, when he wishes to emphasise that they are inherent in matter, *configurations of matter*. He gives repeated lists of simple natures [II, 614; IV, 29, 361; V, 510], a single one of which adequately indicates the sort of property which Bacon regards as a simple nature:

> The *Configurations of Matter* are, Dense, Rare; Heavy, Light; Hot, Cold; Tangible, Pneumatic; Volatile, Fixed; Determinate, Fluid; Moist, Dry; Fat, Crude; Hard, Soft; Fragile, Tensile; Porous, Close; Spiritous, Jejeune; Simple, Compound; Absolute, Imperfectly Mixed; Fibrous and Venous, Simple of Structure, or Equal; Similar, Dissimilar; Specific, Nonspecific; Organic, Inorganic; Animate, Inanimate. [IV, 356]

All these properties (together with the others which occur in alternative lists of simple natures) are regarded by Bacon as being so basic that all kinds of bodies can be completely described in terms of them. What is remarkable about these lists is the uncritical linking of pairs of properties and relations of quite different logical status.

Simple natures are determinate physical properties. They provide an 'alphabet of nature' out of which all the properties of bodies are built:

> For as it would be neither easy nor of any use to inquire the form of the sound which makes any word, since words, by composition and transposition of letters, are infinite; whereas to inquire the form of the sound which makes any simple letter...is comprehensible, nay easy; and yet these forms of letters once known will lead us directly to the forms of words; so in like manner to inquire the form of a lion, of an oak, of gold, nay even of water or air, is a vain pursuit; but to inquire the form of dense, rare, hot, cold, heavy, light, tangible, pneumatic, volatile, fixed, and the like, as well configurations as motions...which (like the letters of the alphabet) are not many and yet make up and sustain the essences and forms of all substances; – this, I say, it is which I am attempting. [IV, 360–1]

(See also [III, 243; IV, 126; V, 426].) Every *compound substance*, that is, every natural kind of body, is defined by some fixed list of simple natures.[1] The effects produced by a compound body may be attributed to the particular collection of simple natures present in it [III, 246].

The *Form* of a simple nature is its essential definition. Bacon apparently envisages this as put together from fundamental terms of the sort which make up the adventitious conditions of

[1] C. D. Broad's identification of simple natures with generic physical properties such as *density*, having as states *dense* and *rare* or the various degrees of density, is slightly misleading. It enables him to relate quantitative scientific laws in the modern sense to Bacon's laws of nature, and this misunderstanding has persisted in the literature. See Broad, *The Philosophy of Francis Bacon* (Cambridge, 1926), reprinted in *Ethics and the History of Philosophy* (London, 1952), and M. B. Hesse, 'Francis Bacon's Philosophy of Science', *A Critical History of Western Philosophy* ed. O'Connor (New York, 1964), reprinted in Vickers ed., *Essential Articles for the study of Francis Bacon* (Connecticut, 1968), pp. 114–39. For Bacon a body of moderate density possesses a mixture of the simple natures *dense* and *rare*, rather than some degree of density [V, 475].

Such lists of pairs of simple or elementary natures are commonplace in writing on natural philosophy in the period, as is their use to define compound substances. The ultimate source for this is Aristotle's *Physics*. See the textbooks cited by P. Reif, 'The textbook tradition in natural philosophy, 1600–1650', *Journal of the History of Ideas*, 30 (1969), pp. 17–32.

essences.[1] The form of any compound body is then the aggregate of the forms of all the simple natures which define it [IV, 122]. Bacon believed that the forms of compound bodies can be made accessible to the investigator only by considering the forms of all the constituent simple natures, because in compound bodies the distinct forms are too much intermingled to be identifiable:

For since every body contains in itself many forms of natures united together in a concrete state, the result is that they severally crush, depress, break, and enthrall one another, and thus the individual forms are obscured. [IV, 158]

The form thus reduces any natural quality to an invariant grouping of fixed essential qualities, inseparable from its occurrence in any kind of matter whatsoever. At the same time, this description in terms of fundamentals is to be a recipe for producing the specified property in any kind of matter whatsoever. Every kind of substance is specified by a particular concomitance of simple natures in matter. Every transformation or modification of a substance is supposed to be equivalent to altering the grouping of simple natures in the body. Gold is defined by the concomitance in matter of *greatness of weight, closeness of parts, fixation, pliantness, immunity from rust, colour or tincture of yellow* [II, 450]. Any matter can be transmuted into gold by operations which induce these simple natures according to the instructions provided by their respective forms:

Gold hath these natures; greatness of weight, closeness of parts, fixation, pliantness or softness, immunity from rust, colour or tincture of yellow. Therefore the sure way (though most about) to make gold, is to know the causes of the several natures before rehearsed, and the axioms concerning the same. For if a man can make a metal that hath all these properties, let men dispute whether it be gold or no. [II, 450]

The forms of simple natures specify the relations between parts of a body which determine a given quality inherent in that body. For example, the form which Bacon postulates for white-

[1] This is nowhere explicitly stated. However, Bacon's two tentative forms (the forms of heat and whiteness) make use of terms like 'part', 'equal', 'unequal', 'motion' which occur among the adventitious conditions of essences.

ness is: *all bodies or parts of bodies which are unequal equally, that is in a simple proportion, do represent whiteness*[1] [III, 237].

Bacon required that the forms of simple natures explain the ways in which change occurs in bodies, and the ways in which some bodies affect others. The form of heat defines heat as a species of motion: *Heat is a motion, expansive, restrained, and acting in its strife upon the smaller particles of bodies* [IV, 154]:

> When I say of Motion that it is as the genus of which heat is a species, I would be understood to mean, not that heat generates motion or that motion generates heat (though both are true in certain cases), but that Heat itself, its essence and quiddity, is Motion and nothing else; limited however by the specific differences which I will presently subjoin. [IV, 150]

Bacon appears to regard heat as particularly important in all transmutation of bodies [IV, 238–40]:

> And it is then that we shall see a real increase in the power of man, when by artificial heats and other agencies the works of nature can be represented in form, perfected in virtue, varied in quantity, and, I may add, accelerated in time. [IV, 240]

We know that the pair *motion, rest* appears in the list of adventitious conditions of essences. Further Bacon believed that in physics the study of compound motions of whole bodies should be broken down into the study of the component simple motions of their parts[2] [IV, 355–7]. It seems reasonable to infer that Bacon believed that the various kinds of motions which occur in the forms of certain privileged simple natures will explain the changes of bodies and their mutual effects in terms of the simple motions of their parts.

If this reconstruction is correct it casts some light on the question raised by many commentators whether Bacon anticipated the primary/secondary qualities distinction. His reduction of simple natures to forms appears to be a reduction of secon-

[1] I shall gloss this definition in a later place (see below, p. 117). For the time being it is only necessary to note that the definition accounts for whiteness in terms of the arrangement of the minute parts of the white body, and in terms of basic relations like 'equally', 'unequally', 'simple proportion'.

[2] Lasswitz notes that the study of motions falls within physics (the study of the variable in nature) rather than within metaphysic. See K. Lasswitz, *Geschichte der Atomistik vom Mittelalter bis Newton* (Hildesheim, 1963) (first edition Hamburg and Leipzig, 1890), vol. I, Book 2, chapter 5, section 4, 'Francis Bacon' (pp. 413–36).

dary qualities to the primary qualities of matter and motion. However, Bacon's wide interpretation of the term 'motion' separates him from any later mechanistic accounts of the primary/secondary qualities distinction. The mechanistic account considers only *local motion*; Bacon includes under the term 'motion' a wide variety of kinds of change. Furthermore, he shows an essentially purposive view of matter by failing to discriminate between change and disposition or inclination to change.[1]

Bacon repeatedly insists that forms are valid only if they are 'grounded in matter'; one of his main objections to previous theories of forms is that both Aristotelian and Platonic notions of 'form' lead to merely arbitrary, nominal definitions, detached from matter. He himself postulates a substratum of homogeneous primary matter, in which forms are grounded:

> And would that this were but agreed on for once by all, that beings are not to be made out of things which have no being; nor principles out of what are not principles; and that a manifest contradiction is not to be admitted. Now an abstract principle is not a being...so that a necessity plainly inevitable drives men's thoughts (if they would be consistent) to the atom; which is a true being, having matter, form, dimension, place, resistance, appetite, motion, and emanations; which likewise, amid the destruction of all natural bodies, remains unshaken and eternal.
>
> [V, 492]

Bacon's requirement that there should exist a substratum of homogeneous matter may cast some light on the apparent inconsistencies in his pronouncements on atomism. It has been pointed out that whilst Bacon often expresses admiration for Democritus and Democritean atomism he in fact in one place or another rejects or questions every specific tenet of atomism except the postulation of the homogeneity of atoms.[2] And an apparent evolution of Bacon's views on atomism has been traced from an early enthusiasm for Democritean atomism as a physical hypothesis to his rejection of atomism in the *Novum Organum* as a physical hypothesis in favour of an explanation of physical

[1] See M. Primack, 'Outline of a reinterpretation of Francis Bacon's philosophy', *Journal of the History of Philosophy*, 5 (1967), pp. 123–32. Compare Bacon's attitude with those of contemporary textbook writers described by Reif, art. cit., pp. 26–7.

[2] See M. Primack, *Francis Bacon's Philosophy of Nature, and Teleology and Mechanism in the Philosophy of Francis Bacon* (Unpublished Ph.D. diss., Johns Hopkins, 1962), pp. 51–136.

change in terms of tenuous spirits.[1] I suggest that Bacon's vacillation on the question of atomism may arise in part from his finding in atomism a convenient metaphysical doctrine which guarantees a homogeneous substratum in which forms inhere, whilst at the same time finding atomism as a physical hypothesis incompatible with his strictly empirical requirements for explanation.

Operative and contemplative knowledge

By allowing kinds of motion to enter into definitions of forms Bacon is able to envisage forms as providing instructions for producing change. This is what he calls the 'operative' aspect of forms.

Bacon argues that there can be only one purpose for human knowledge, and that is the power to act [IV, 47, 119]. Definitions which fail to enlarge the scope of man's practical activities cannot contribute to real knowledge. He therefore requires that the forms of simple natures arrived at by way of his own inductive method should be guaranteed to produce fruitful consequences. This he hopes to achieve by matching his method to the stages of human cognition. However abstract the intermediate stages of the method may appear, it is a tool matched both to nature and to man's capacities to know, and its results will be not merely academically interesting, but of immediate practical consequence [IV, 107]:

> But whosoever is acquainted with Forms, embraces the unity of nature in substances the most unlike; and is able therefore to detect and bring to light things never yet done, and such as neither the vicissitudes of nature, nor industry in experimenting, nor accident itself, would ever have brought into act, and which would never have occurred to the thought of man. From the discovery of Forms therefore results truth in speculation and freedom in operation. [IV, 120]

The requirement that forms of simple natures shall be at once essential definitions of natural qualities, and *laws* or practical rules for producing such qualities is the key to Bacon's whole programme for replacing the scholastic division of science into

[1] Lasswitz, op. cit. See also M. Macciò, 'A proposito dell' atomismo nel "Novum Organum" di Bacone', *Rivista critica di storia della filosofia*, 17 (1962), pp. 188–96. Macciò traces Bacon's changing views to changes in his physical beliefs.

operative (or productive) and contemplative by a unified natural philosophy. In Bacon's scheme operative and contemplative knowledge are indissolubly linked:

For when I speak of Forms, I mean nothing more than those laws and determinations of absolute actuality, which govern and constitute any simple nature, as heat, light, weight, in every kind of matter and subject that is susceptible of them. Thus the Form of Heat or the Form of Light is the same thing as the Law of Heat or the Law of Light. Nor indeed do I ever allow myself to be drawn away from things themselves and the operative part. And therefore when I say (for instance) in the investigation of the form of heat, 'reject rarity,' or 'rarity does not belong to the form of heat,' it is the same as if I said, 'It is possible to superinduce heat on a dense body;' or, 'It is possible to take away or keep out heat from a rare body.' [IV, 146]

Bacon's main reason for stipulating this double character of forms is that only if the forms are practically applicable can their truth be tested by the practical consequences it produces. Bacon's insistence on the practical efficacy of his forms is not an indication that he was predominately concerned with the growth of technology. Bacon's point is that only through practical consequences can the truth of a postulated scientific explanation ever be verified:[1]

But I say that those foolish and apish images of worlds which the fancies of men have created in philosophical systems, must be utterly scattered to the winds. Be it known then how vast a difference there is (as I said above) between the Idols of the human mind and the Ideas of the divine. The former are nothing more than arbitrary abstractions; the latter are the creator's own stamp upon creation, impressed and defined in matter by true and exquisite lines. Truth therefore and utility are here the very same things [Itaque ipsissimae res sunt (in hoc genere) veritas et utilitas]: and works themselves are of greater value as pledges of truth than as contributing to the comforts of life. [IV, 110]

A self-consistent system may be a mere fantasy, in no way matching natural order (the *idea* in the divine mind). Works are

[1] A number of critics have claimed Bacon as the patron of technology, notably B. Farrington, *Francis Bacon: Philosopher of Industrial Science* (New York, 1949); C. A. Viano, 'Esperienza e Natura nella filosofia di Francesco Bacone', *Rivista di Filosofia* (1954), pp. 291–313; F. Anderson, *The Philosophy of Francis Bacon* (Chicago, 1948); R. C. Cochrane, 'Francis Bacon and the use of the mechanical arts in eighteenth century England', *Annals of Science*, 12 (1956), pp. 137–56; M. E. Prior, 'Bacon's Man of Science', *Journal of the History of Ideas*, 15 (1954), pp. 340–70, reprinted in Vickers ed., *Essential Articles for the study of Francis Bacon* (Connecticut, 1968), pp. 140–63.

the test of the correspondence between the *principia* of the proposed system, and the facts of nature, and hence there can be no real distinction made between the 'truth' and the 'works' which guarantee that truth.[1] In fact, contemplative truth, not technological advance is the main goal of Baconian science:[2]

Again it will be thought, no doubt, that the goal and mark of knowledge which I myself set up (the very point which I object to in others) is not the true or the best; for that the contemplation of truth is a thing worthier and loftier than all utility and magnitude of works; and that this long and anxious dwelling with experience and matter and the fluctuations of individual things, drags down the mind to earth, or rather sinks it to a very Tartarus of turmoil and confusion; removing and withdrawing it from the serene tranquillity of abstract wisdom, a condition far more heavenly. Now to this I readily assent; and indeed this which they point at as so much to be preferred, is the very thing of all others which I am about. [IV, 110]

For Bacon the 'operative' aspect of his forms of simple natures is valuable chiefly for its ability to provide verification of the form itself, and only secondarily for the guidance it provides for technology:

For I care little about the mechanical arts themselves: only about those things which they contribute to the equipment of philosophy. [IV, 271]

On the face of it it is hard to see how a definition could be equivalent to a formula for successfully doing something. It is therefore necessary to examine the way in which Bacon manages to sustain this equivalence to his own satisfaction.

In the *Valerius Terminus* Bacon gives the form of whiteness as follows: *all bodies or parts of bodies which are unequal equally, that is*

[1] Rossi devotes an appendix in *I filosofi e le macchine (1400–1700)* (Milan, 1962), pp. 148–73, to an extended discussion of Bacon's use of the terms 'veritas' and 'utilitas', centring on the interpretation of the passage: 'itaque ipsissimae res sunt (in hoc genere) veritas et utilitas' in the passage cited above. He replaces Spedding's translation (about which Spedding himself expresses misgivings in a footnote) by the following: 'le cose, cosí come esse realmente sono, considerate non del punto di vista dell'apparanza, ma da quello dell'esistenza, non in rapporto all'uomo, ma in rapporto all'universo, offrono congiuntamente la verità e l'utilità.' Rossi presents clearly the textual evidence for maintaining, as I do, that Bacon believed his method by its very definition to elide contemplative truth and operative effectiveness. However, Rossi's conclusion does not put this very usefully: 'ricerca teorica e applicazione pratica non sono che *la stessa esperienza che si configura in due modi diversi.*'

[2] This point is recognised by A. R. Hall. See Hall, *The Scientific Revolution 1500–1800* (London, 1954), pp. 164–9.

in a simple proportion, do represent whiteness [III, 237]. That is, whiteness and opacity are present where the smallest parts of any body are of unequal size, and are intermingled in some fixed ratio or proportion. Further in the same paragraph he adds:

> It is then to be understood, that absolute equality produceth transparence, inequality in simple order or proportion produceth whiteness, inequality in compound or respective order or proportion produceth all other colours, and absolute or orderless inequality produceth blackness. [III, 237]

That is, where the parts of a body are equal, transparency is *produced*; where the smallest parts of a body are all unequal, and are randomly mixed together, blackness in the body is *produced.* Whilst in the third Book of the *De Augmentis* the form of whiteness appears as follows: *two transparent bodies intermixed, with their optical portions arranged in a simple and regular order, constitute whiteness* [IV, 361].

In the *Novum Organum* Bacon arrives at the following definition of heat:

> Heat is a motion, expansive, restrained, and acting in its strife upon the smaller particles of bodies...while it expands all ways, it has at the same time an inclination upwards, [the struggle] is not sluggish, but hurried and with violence. [IV, 154–5]

This definition Bacon immediately restates as equivalent to the rule of operation:

> If in any natural body you can excite a dilating or expanding motion, and can so repress this motion and turn it back upon itself, that the dilation shall not proceed equably, but have its way in one part and be counteracted in another, you will undoubtedly generate heat. [IV, 155]

In both examples Bacon inadvertently slides from stating the form as a definition, '*x* is the form of *y*', to stating it as a rule of operation, '*x* invariably produces *y*'. And this sustains his belief that each of the forms of simple natures is at once a definition and a rule of operation – is at once a source of contemplative and operative knowledge.

Forms as the basis for works

Forms are the culmination of the 'ascent' from works to principles, guided by the inductive method; what follows is the 'descent' to further works (see above, p. 99). Bacon nowhere discusses the method by which this descent is to be carried out, although his endorsement of syllogistic once first principles are secure (see above, p. 86) leads one to believe that this is the form of inference to be used for this stage in the proceedings. From the forms all physical generalisations are to be deduced (including the local generalisations provisionally arrived at in physics), and hence unlimited control over natural phenomena will be made available.

One may, rather tentatively, reconstruct on the basis of Bacon's own data the kind of procedure which he envisaged as leading from forms to explanations of observed natural processes. For instance, Bacon gives two preliminary definitions of simple natures in his works:

> Heat is a motion, expansive, restrained, and acting in its strife upon the smaller particles of bodies. [IV, 154]

> All bodies or parts of bodies which are unequal equally, that is in a simple proportion, do represent whiteness. [III, 237]

He therefore presumably imagined that a body which is both white and hot (for instance, hot milk) will have as part of its composite form an expansive, restrained motion of its smaller parts, and the parts arranged in simple proportion. On the basis of these components of the definition of the body Bacon apparently expects to be able to deduce the subsequent behaviour of the body as a whole. For instance he would perhaps expect hot milk to boil over on the following grounds: the expansive motion of parts of a body which are of unequal size will produce uneven motion. 'While [this motion] expands all ways, it has at the same time an inclination upwards' [IV, 155]. Because the unequal parts are arranged in a simple proportion or ratio within the body, the irregular upwards motion will predominate. Hence the milk will overflow the pan.

Although the above is my own contrived example, Bacon's explanation of the motion of a projectile in terms of the definitions of its component motions (arising from its form) follows the same pattern:

But the most evident effect of this motion [compression] is seen in the perpetual revolutions or rotations of projectiles in their flight...the cause of this effect is the same that I am now speaking of, and no other. For pressure of a body at once excites a motion in the parts or particles to extricate and free themselves in any way they can. And hence the body is not only driven in a straight line, and so flies forward; but it tries all round, and therefore revolves; for both motions help to set it free. [V, 435]

And likewise the true motions and orbits of the planets are to be derived from their component forms:

from those primary and catholic axioms concerning simple natures; such as the nature of spontaneous rotation, of attraction or magnetism, and of many others which are of a more general form than the heavenly bodies themselves. [IV, 123]

In all particular bodies, under particular circumstances, the changes and effects produced are to be derived by compounding configurations and motions which are more universal, that is 'better known in nature'.

6

The interpretation of nature

We are now in a position to consider the detailed procedure of Bacon's proposed method for interpreting nature. This method is to enable the investigator to arrive at true definitions of natural kinds, starting from basic sense-experience, by simple stages. The definitions are to have the special virtue of being at once the basis for subsequent logical deduction, and instructions for successful practical operations.

Certainty and freedom

Aristotle had stipulated that definitions which are to serve as the basis for deduction of scientific knowledge, or demonstration, must fulfil three criteria. Such *principia* must be true *universally*, *primarily*, and *essentially*.[1] A definition is true universally if the predicate is true of the subject in every instance.[2] It is true essentially if the predicate is part of the essential nature of the subject, or if it is part of the cause of the subject's occurrence (causes are said to be true essentially of their effects because for Aristotle the formal cause and essential definition of a thing are identical).[3] A definition is true primarily if the predicate is true of each member of the class of things denoted by the subject, and that class is the widest to which the predicate can be applied.[4]

[1] *Posterior Analytics* I, 4. In *Topics* VI these criteria occur again, as part of the detailed procedure for checking an adversary's definitions at the beginning of a disputation.

[2] The proposition 'all men are mortal' is true universally, since whatever is a man is automatically mortal.

[3] *Odd* and *even* are true essentially of *number*, since any number is of its essential nature odd or even.

[4] Aristotle's example is that the predicate 'angles equal to two right angles' is true primarily of the subject *triangle*, because it is true for every triangle, but is not true of any class of

These criteria for the axioms of sciences amount to the requirements that the definitions on which deductive reasoning in the sciences is based should be such that a definition is true in every instance of the subject's occurrence, is essential to it as opposed to accidental, and defines no class more inclusive than the subject. Given such axioms, all subsequent reasoning using syllogistic will yield conclusions whose truth is allegedly assured.

Bacon accepts Aristotle's three criteria as necessary for the first principles of the sciences [IV, 453]. His inductive method is designed to yield definitions fulfilling these criteria on the basis of strictly empirical evidence. He apparently believed that essential truth is guaranteed by the initial procedure of intuitive selection of simple natures and adventitious conditions of essences from the preliminary natural history. That is, he believed that it is immediately obvious to the experienced investigator which properties and relational terms are most widely employed in all accounts of natural processes, and hence form the basis for true definitions of natural kinds.

He introduces the remaining two rules as the rules of *certainty* and *freedom*. It is required that the forms (or laws of nature) shall be such that whenever the conditions of the definition are fulfilled the simple nature defined is inevitably produced (the rule is *certain*), and whenever the conditions are not fulfilled the simple nature is inevitably absent (the rule is *free*).

If therefore your direction be certain, it must refer you and point you to somewhat which, if it be present, the effect you seek will of necessity follow, else may you perform and not obtain [if it is *not* certain performing the operation may not produce the required effect]. If it be free, then must it refer you to somewhat which if it be absent the effect you seek will of necessity withdraw, else may you have power and not attempt [if it is *not* free one may be able to produce the effect under some circumstances but not at will]. This notion Aristotle had in light, though not in use. [III, 235–6][1]

figures larger (more inclusive) than triangles. 'Angles equal to two right angles' is not true primarily of *isosceles triangle*, since *triangle* is a prior specification; nor is it true primarily of *figure*, since the square (for instance) has angles equal to four right angles. Similarly, 'suckle their young with their milk' is true primarily of *mammals*, since it is true for each of the class of mammals, and is not true of any more inclusive class. It is not true primarily of *rodents*, since the predicate is true of the more inclusive class *mammals*.

Caesarius treats these three rules, which he calls the rules of truth *de omni*, *per se* and *u):versale*, under *demonstration*. *Dialectica* (1568 ed.), fo. S5r-S7v.

[1] Bacon's description of the procedure for ensuring certainty and freedom of definitions in the *Valerius Terminus* is strikingly reminiscent of Aristotle in *Topics* VI on checking definitions.

Since Bacon believed that his Forms will at once provide essential definitions of simple natures, and recipes for their practical production, he apparently supposed that the criteria of certainty and freedom extend both to the logical status of the definitions obtained by the inductive method, and to the effectiveness of the corresponding rules of operation for producing simple natures:

For a true and perfect rule of operation then the direction will be *that it be certain, free, and disposing or leading to action.* And this is the same thing with the discovery of the true Form. For the Form of a nature is such, that given the Form the nature infallibly follows. Therefore it is always present when the nature is present, and universally implies it, and is constantly inherent in it. Again, the Form is such, that if it be taken away the nature infallibly vanishes. Therefore it is always absent when the nature is absent, and implies its absence, and inheres in nothing else. Lastly, the true Form is such that it deduces the given nature from some source of being which is inherent in more natures, and which is better known in the natural order of things than the Form itself.[1] For a true and perfect axiom of knowledge then the direction and precept will be, *that another nature be discovered which is convertible with the given nature, and yet is a limitation of a more general nature, as of a true and real genus.* Now these two directions, the one active the other contemplative, are one and the same thing; and what in operation is most useful, that in knowledge is most true. [IV, 121–2]

Conversely, a method which derives certain instructions for the production of selected simple natures under all circumstances will automatically satisfy the requisite logical criteria for definitions of first principles.

In practice Bacon attempts to use an analysis of the means of production of simple natures to provide scientific definitions fulfilling Aristotle's criteria of essential, primary and universal truth:

Although the roads to human power and to human knowledge lie close together, and are nearly the same, nevertheless on account of the pernicious and inveterate habit of dwelling on abstractions, it is safer to begin and raise the sciences from those foundations which have relation to practice, and to let the active part itself be as the seal which prints and determines the contemplative counterpart. [IV, 120–1]

[1] This last direction gives the condition for the definition being *essentially* true (see Aristotle, *Topics* VI, 4 (141a22-141b2)). Certainty and freedom ensure that the form is convertible, this last condition ensures that it is couched in terms which are 'better known in nature' than the simple nature itself.

In his discussion of certainty and freedom he concentrates his attention entirely on the operative aspect of the forms. In this context the criteria of certainty and freedom correspond to sufficient and necessary conditions respectively for the production of a simple nature.[1] The procedure of the inductive method is designed to give necessary conditions for the simple nature's occurrence by elimination of extraneous instances from the complete natural history of telling instances of a specified nature. These conditions are then strengthened so as to ensure that they are also sufficient.

Preliminary stages in the interpretation of nature

The basis for the inductive method is direct sense-experience. According to Bacon's faculty psychology this is recorded and stored in primitive form in the memory. In written form it provides the elementary natural history on which all natural philosophy is based. The investigator's first move, at the outset of the methodical investigation of a simple nature, is to collect from the complete natural history a subsidiary history comprising every occurrence of the simple nature he has chosen to investigate [III, 552].

From this history the investigator picks out all occurrences of the simple nature which are particularly informative. He selects instances of the simple nature which are likely to provide useful shortcuts to the discovery of its form. I shall return to the details of this selection shortly. These privileged instances are collated and organised so as to provide as clear as possible a picture of the production of the selected simple nature. Where the investigator can imagine possible but untried experiments which will yield further information, or which fill gaps in the existing history, he designs and carries out such experiments, and records their results.[2] When this tabulation is completed the investigator is in a position to use an induction by elimination to arrive at necessary conditions for the nature's production, which are then adjusted for sufficiency. The resulting rule of operation

[1] This is pointed out by C. J. Ducasse, 'Francis Bacon's Philosophy of Science', Henle *et al.* ed., *Structure, Method and Meaning* (New York, 1951), pp. 115–44.

[2] This procedure of designing and performing new experiments Bacon calls *Experientia Literata*. See below, chapter 7.

simultaneously *defines* the simple nature, and can be used as the basis for further deduction.[1]

Prerogative instances

Bacon devotes the major part of the discussion of the interpretation of nature in the *Novum Organum* to specific instructions for ensuring that the best possible use is made of the senses and memory in assembling the material for the natural history, and for tabulating its contents.

Amongst the instances of a simple nature recorded in the history there will always be certain outstanding instances which in one way or another make it particularly clear how the nature is put together, or produced. Such instances may be used to provide shortcuts in the investigation, and Bacon assigns to them a special privilege or *prerogative* [III, 556; IV, 258]. Prerogative instances are of two basic kinds. Some give practical help in devising experiments, extending and magnifying the power of the senses in observing natural phenomena, and suggesting useful practical ways of harnessing simple natures for technological benefit. Some, on the other hand, are of *theoretical* importance, and provide natural justification for decisions about the treatment of data. They solve problems of procedure in the course of the organisation of sense-experience prior to the inductive derivation of the form of a simple nature. In particular, where two conflicting courses of action are open to the investigator, or two directions in which the inquiry may proceed, a prerogative instance which directly refutes one alternative decides the issue.

Some prerogative instances aid the senses directly, by indicating special circumstances under which the senses are used to their best advantage, and the way in which technical aids can be used to record effects inaccessible to the unaided sense. Bacon groups such prerogative instances together as *Instances of the Lamp* [IV, 192, 246]. The instance which records sunspots, seen through a telescope has a prerogative of this sort [IV, 193]. In investigating the form of the sun, such an instance would rank

[1] A clear general account of the stages of the interpretation of nature which precede the induction itself occurs in [III, 552–7]. The *Novum Organum* gives a worked example, the investigation of the form of heat [IV, 126–248].

above those involving the unaided senses, and would occupy a more prominent position in the tabulation.

Other instances narrow down the field of inquiry, by indicating effects and properties which invariably accompany the occurrence of the simple nature in question, and which hence figure in the form of that nature, or by excluding properties not essential to the simple nature, or effects wrongly considered the product of a nature. A *Migratory Instance* has a prerogative of this kind [IV, 156, 247]. A migratory instance is one in which the nature under investigation appears, disappears, increases or diminishes. By noting the characteristics of the situation which alter with the growth or diminution of the nature in question the investigator is able to assert that such characteristics are invariably linked with the nature. This immediately restricts the area of investigation. Thus if whiteness is produced by pounding transparent glass to a powder this tells the investigator (according to Bacon) that air and pounding are responsible for the transformation of transparency into whiteness, since nothing else was added to the glass during the pounding [IV, 157].

An *Instance of the Fingerpost* (also called a *Decisive and Judicial Instance*) [IV, 180, 247] presents the nature under investigation in such a way as to decide between two alternative explanations. For example, if the question is whether all heavenly bodies move round the earth from west to east, then an instance of some heavenly body, such as a comet, observed to move through the sky from east to west refutes the hypothesis [IV, 184].

These prerogative instances contribute to the 'informative part' of the investigation of nature – the quest for forms [IV, 246]. Prerogative instances are also of direct help to the operative part [IV, 246]. They give direct information about the possibilities for exploiting natural processes, without recourse to forms. Instances of this sort belong strictly to technology. For instance, amongst *Deviating Instances*, or instances which record aberrations in nature, monsters and prodigies, it may be possible to find ways of modifying nature which will (say) increase productivity. If, under extraordinary circumstances, wheat gives twice its usual yield, it may be possible to recreate those circumstances artificially, and so ensure an increased harvest [IV, 168–9, 247].

The prerogative instances fill out with empirical advice Bacon's largely academic account of the stages of his inductive method. The discussion of prerogative instances occupies a very large part of the account of the inductive method in the *Novum Organum*. Most of the interesting observations which Bacon makes about scientific procedure are to be found amongst his remarks on specific instances with special prerogative.[1]

The induction by elimination

With the help of the prerogative instances the investigator is able to reduce the instances of a given simple nature to be examined to a small number of significant ones which manifest that nature in particularly suggestive ways. These form the *Table of Presence*. A second table matches each instance in the first with an instance of absence of the nature, under circumstances which differ in as few respects as possible from those in the first table. These tables (a *Table of Presence* and a *Table of Absence*) are supplemented by a *Table of Variation*, in which instances are listed where the increase or decrease in the nature in question is accompanied by increase or decrease in other natures present, indicating that these may be essential concomitants of the nature under investigation.[2] Elimination between the table of presence and the table of absence provides the final stage in the inductive method.

For Bacon this simple elimination between the tables is the only natural and legitimate use of formal inference in the interpretation of nature. Any more complicated rational activity will, he claims, be too far removed from immediate experience, and hence will inevitably be distorted by the various *idols*. A simple elimination, however, at the end of a series of natural

[1] Further 'prerogatives' will, according to Bacon, solve all the procedural problems in the later stages of the inductive method – in what order to consider simple natures, how to check the induction, how to apply results, and so on [IV, 155; III, 556]. The *Novum Organum* is incomplete in the sense that Bacon fails to make explicit the sort of use to which such aids to the investigation will be put. He evidently considered such questions to be largely a matter of experience, familiarity with natural processes, and astute selection of appropriately informative instances. The important stages of the investigation, however, (the tabulation and induction) are presented in complete form in the *Novum Organum*. C. D. Broad nevertheless suggests that the method would be more easily comprehensible if only vital missing portions had been completed. See Broad, *The Philosophy of Francis Bacon* (Cambridge, 1926), reprinted in *Ethics and the History of Philosophy* (London, 1952), pp. 117–43. Bacon claimed that he had ready in his own mind a final book for the *Novum Organum* [II, 336].

[2] Aristotle gives concomitant variation as a way of checking a definition in a disputation, *Topics* VI, 7 (146a2-12).

manipulations of raw sense data, provides an interpretation of nature which is 'the true and natural work of the mind when freed from impediments' [IV, 115]:

For I am of opinion that if men had ready at hand a just history of nature and experience, and laboured diligently thereon; and if they could bind themselves to two rules, – the first, to lay aside received opinions and notions; and the second, to refrain the mind for a time from the highest generalisations, and those next to them, – they would be able by the native and genuine force of the mind, without any other art, to fall into my form of interpretation. For interpretation is the true and natural work of the mind when freed from impediments. [IV, 115]

Indulgences of the understanding

Only if the list of simple natures is correct and the history is complete can the simple elimination produce the final form of the simple nature under investigation [IV, 149].[1] Failing a complete list of simple natures and complete history, the elimination provides the basis for an *Indulgence of the Understanding* or *Commencement of Interpretation* [IV, 149]. A form consistent with the result of the elimination is postulated, and provides the first stage, or *First Vintage* in the interpretation.[2] The provisional form must be tested to ensure that it fits all possible occurrences of the simple nature, and does not cover occurrences of any other simple nature. That is, it must be tested to ensure that it provides necessary and sufficient conditions for the production of the simple nature. In Bacon's two worked examples of his

[1] Several critics have recognised this feature of the induction, without taking into account the fact that Bacon in practice adapts his method to the inevitable insufficiency of the data (although in the long run he probably hoped that a perfect induction would be possible). See for instance, Broad, op. cit.; M. B. Hesse, 'Francis Bacon's Philosophy of Science', O'Connor ed., *A Critical History of Western Philosophy* (New York, 1964), reprinted in Vickers ed., *Essential Articles for the study of Francis Bacon* (Connecticut, 1968), pp. 114–39; H. W. Blunt, 'Bacon's method of science', *Proceedings of the Aristotelian Society*, 4 (1903–4), pp. 16–31.

[2] Bacon proposed to devote the fourth part of the Great Instauration to examples of his method in this provisional state: 'The first [thing] is to set forth examples of inquiry and invention according to my method, exhibited by anticipation in some particular subjects; choosing such subjects as are at once the most noble in themselves among those under inquiry, and most different one from another; that there may be an example in every kind. I do not speak of those examples which are joined to the several precepts and rules by way of illustration (for of these I have given plenty in the second part of the work) [the examples in the *Novum Organum*]; but I mean actual types and models, by which the entire process of the mind and the whole fabric and order of invention from the beginning to the end...should be set as it were before the eyes' [IV, 31].

inductive method, the heat example in the *Novum Organum* and the whiteness example in the *Valerius Terminus*, the provisional form gives necessary but not sufficient conditions for the occurrence of the simple nature in question. Although the form covers all instances cited of the simple nature in question it also covers instances of some closely related natures, and has to be modified accordingly [III, 237; IV, 151].

Bacon allows a cautious guess at a form, based on a careful tabulation of the contents of the history for the simple nature concerned because of his belief that his method, closely tied as it is to experience, enables the investigator to go back at every stage and check or modify earlier stages in the interpretation. The natural history is corrected in the light of later discoveries [IV, 260]. Simple natures may be adjusted in the course of investigation [IV, 149, 161–2]. Even the method may be modified as knowledge of simple natures grows [IV, 115]. Guesses within the framework of the method do not, therefore, suffer from the same defects as the 'anticipations of the mind' [IV, 42] which according to Bacon marred traditional investigations.

Bacon's own examples of interpretation of nature

We have two examples in Bacon's writings of attempts to use the inductive method. The first (and earliest) is the investigation of the form of whiteness in the *Valerius Terminus*; the second is the investigation of the form of heat in the *Novum Organum*.

The first stage in the investigation of the form of whiteness is the selection of instances of the production of whiteness which provide certain (sufficient) conditions for the production of whiteness in all bodies. Air and water intermingled produce whiteness; air and glass intermingled produce whiteness. Both these 'migratory instances' [IV, 157] give manifest material and efficient causes of whiteness. Both tie the production of whiteness to particular physical substances, namely to air and water in the first case, and to air and glass in the second. Since the same effect is produced in each instance, this suggests (by eliminating all that is not common to the two occurrences) that air and any transparent body compounded make whiteness. Further instances of the production of whiteness allow further generalisation of the instruction until the formulation is reached:

'that if any bodies, both transparent but in unequal degree, be mingled...whiteness will follow.' This formulation *frees* the form of whiteness from the material constraints of the particular instances of it which are examined. Bacon goes on to suggest provisionally a further freeing of the form to give: 'all bodies or parts of bodies which are unequal equally, that is in a simple proportion, do represent whiteness'.[1] This is supposed to be equivalent to the rule of operation: 'unequal parts combined in a simple proportion produce whiteness.' The provisional form is only to be accepted after ratification by examination of the history of instances of 'colour and visible objects' [III, 235–7].

But the final freeing of whiteness which results from this formulation needs to be amended for certainty. Extra conditions have to be imposed to get round the fact that various instances of mixing of parts of bodies in simple proportion can be produced in which whiteness is *not* produced. This is done by specifying that the parts thus intermingled must be *large enough* if whiteness is to ensue. The form of whiteness finally arrived at in this way is supposed to be both a definition of whiteness in terms of fundamentals like 'unequal', 'equal', 'part', 'simple proportion', and an instruction for producing whiteness under all circumstances. Simple natures whose forms contain related lists of fundamental terms apparently belong to the same family of qualities:

> It is then to be understood, that absolute equality produceth transparence, inequality in simple order or proportion produceth whiteness, inequality in compound or respective order produceth all other colours, and absolute or orderless inequality produceth blackness.
>
> [III, 237]

(See also [IV, 157–8].) The family of colours, therefore, all have related forms.

It should be noted that the inductive elimination is particularly simple in this case because prerogative *migratory* instances are selected from amongst all possible instances of occurrence of whiteness. The area of investigation is narrowed to precisely those features which are altered by the production of whiteness.

[1] For a gloss of this definition see above, p. 117.

Whiteness is immediately narrowed to a combination of 'air', 'transparent' and 'compounded'.[1]

In the heat example in the *Novum Organum* a more representative selection of instances is made to form the table of presence. The instances include:

The rays of the sun, especially in summer and at noon.
All flame.
Green and moist vegetables confined and bruised together, as roses packed in baskets; insomuch that hay, if damp when stacked, often catches fire.
Quick lime sprinkled with water.
Oil of marjoram and similar oils have the effect of heat in burning the bones of the teeth. [IV, 127–8]

All these instances are offered as causes of the effect heat, or as instances which produce heat. It is of course noticeable that Bacon fails to distinguish between instances where heat is produced, and instances where a 'hot' sensation is produced. A corresponding list is made of allied instances in which heat is not produced. For example, the instance of the rays of the moon is opposed to that of the rays of the sun, since in the former case no heat is produced.

A table is also constructed of *Degrees or Comparison in Heat.* This records the differences in intensity of heat in different instances, and instances in which heat increases or decreases. For instance:

Animals increase in heat by motion and exercise, wine, feasting, venus, burning fevers, and pain.
Some ignited bodies are found to be much hotter than some flames. Ignited iron, for instance, is much hotter and more consuming than flame of spirit of wine.
An anvil grows very hot under the hammer, insomuch that if it were made of a thin plate it might, I suppose, with strong and continuous blows of the hammer, grow red like ignited iron. [IV, 139, 142]

In all three tables Bacon indicates how the tables are to be enlarged by carrying out new experiments. Such experiments

[1] In an otherwise correct account of the logical form of Bacon's inductive method, M. B. Hesse suggests that there are successive levels of forms, and that after this tentative form of whiteness has been discovered a further induction will obtain this form in terms of yet more fundamental qualities. However, as a result of a *single* induction Bacon apparently believed that it would be immediately evident which simple natures fall into groups by virtue of the shared fundamental terms in their forms. See Hesse, art. cit.

are on the whole suggested by others which have produced successful results in comparable circumstances, or as opposite techniques, which may produce contrary results. For instance, the following is suggested in the table of absence:

> Try the following experiment. Take a glass fashioned in a contrary manner to a common burning-glass, and placing it between your hand and the rays of the sun, observe whether it diminishes the heat of the sun, as a burning-glass increases and strengthens it. For it is evident in the case of optical rays that according as the glass is made thicker or thinner in the middle as compared with the sides, so do the objects seen through it appear more spread or more contracted. Observe therefore whether the same is the case with heat. [IV, 131]

Such experiments form a part of *Experientia Literata*, the systematisation and extension of experience by analogy with past experiences [IV, 413–21] (see below, chapter 7). *Experientia Literata* provides both the systematic observation needed as a basis for interpretation of nature, and an independent, less rigorous way of devising new operations on the basis of old ones [IV, 96].

The tables form the basis for the *exclusion*. All natures which are not common to all instances of presence of heat, or which do not disappear in the absence of heat, or do not vary with variations in heat, are eliminated:

> On account of the rays of the sun, reject the nature of the elements.
> On account of air, which is found for the most part cold and yet remains rare, also reject rarity.
> On account of ignited iron, which does not swell in bulk, but keeps within the same visible dimensions, reject local or expansive motion of the body as a whole. [IV, 147–8]

On the strength of the exclusion Bacon once again suggests a provisional form as the 'commencement of interpretation':

> From a survey of the instances, all and each, the nature of which Heat is a particular case appears to be Motion. This is displayed most conspicuously in flame, which is always in motion, and in boiling or simmering liquids, which also are in perpetual motion. It is also shown in the excitement or increase of heat caused by motion, as in bellows and blasts...Again it is shown in the extinction of fire and heat by any strong compression, which checks and stops the motion...It is shown also by this, that all bodies are destroyed, or at any rate notably altered, by all

strong and vehement fire and heat; whence it is quite clear that heat causes a tumult and confusion and violent motion in the internal parts of a body, which perceptibly tends to its dissolution. [IV, 150]

Again, restrictions have to be placed on the definition of heat as *violent motion*, to cover the fact that some violent motion produces natures other than heat. The final form of heat arrived at as a result of these restrictions is:

Heat is a motion, expansive, restrained, and acting in its strife upon the smaller particles of bodies. But the expansion is thus modified; *while it expands all ways, it has at the same time an inclination upwards.* And the struggle in the particles is modified also; *it is not sluggish, but hurried and with violence.* [IV, 154–5]

The rule of operation corresponding to this form is:

If in any natural body you can excite a dilating or expanding motion, and can so repress this motion and turn it back upon itself, that the dilation shall not proceed equably, but have its way in one part and be counteracted in another, you will undoubtedly generate heat. [IV, 155][1]

[1] In both these examples of the inductive method in action Bacon regards *combinations* of properties as simple aggregates without interaction. In the whiteness example, 'intermingling' and 'pounding together' are identified with 'compounding', so that in the final form the term 'combination' can be taken as self-explanatory.

For accounts of the logic of the induction and its relation to the inductive methods of Herschel and Mill see C. von Sigwart, *Logik* (Tübingen, 1873), transl. Dendy (London, 1890), vol. II, chapter 5; Blunt, art. cit.; von Sigwart points out that the form of inference used in the induction is strictly syllogistic.

7

Analogy and generalisation in natural philosophy

The goal of Bacon's inductive method is definitions which are at once self-evidently true and practically effective. But the Organon by means of which Bacon hoped to achieve this goal remained, on his own admission, unperfected [IV, 91, 104, 115]. There is, however, no question of suspending the investigation of nature until the Organon is perfected, and forms are discovered. In expectation that the true interpretation of nature will eventually be effected, less general investigations are to be carried out in the meantime, whose results will at least enable men to begin harnessing the forces of nature:

> These [less general] inquiries also relate to natures concrete or combined into one structure, and have regard to what may be called particular and special habits of nature, not to her fundamental and universal laws which constitute Forms. And yet it must be confessed that this plan appears to be readier and to lie nearer at hand and to give more ground for hope than the primary one. [IV, 123]

The generalisations arrived at by means of the less ambitious form of investigation will nevertheless allow men to begin perfecting mechanics (the applied study corresponding to physics in Bacon's classification of knowledge). Progress in natural magic (the applied study corresponding to metaphysic) must, however, wait until forms, and the first principles of knowledge have been discovered:

> The operative which answers to this speculative part [the less general study of nature], starting from the ordinary incidents of nature, extends its operation to things immediately adjoining, or at least not far removed. But as for any profound and radical operations on nature, they depend entirely on the primary axioms. [IV, 123]

Although Bacon presents the investigation of local, less general causes and principles as a parallel study to the investigation of forms [IV, 122–3], he presumably intended this study to be subsumed under metaphysic once the true interpretation of nature is completed. The metaphysics of forms will account for all the provisional regularities discovered by preliminary excursions into the investigation of natural phenomena.

From his own investigations Bacon is aware that, in the course of assembling material for his natural histories, the first stage of the inductive method, observation of regularities and recurrences will lead the investigator to make tentative generalisations about the phenomena. Although he stresses their provisional nature, he accepts such generalisations as useful, and intends the fifth part of the Great Instauration to comprise a collection of them:

> I include in this fifth part such things as I have myself discovered, proved, or added, – not however according to the true rules and methods of interpretation, but by the ordinary use of the understanding in inquiring and discovering. For besides that I hope my speculations may in virtue of my continual conversancy with nature have a value beyond the pretensions of my wit, they will serve in the meantime for wayside inns, in which the mind may rest and refresh itself on its journey to more certain conclusions. Nevertheless I wish it to be understood in the meantime that they are conclusions by which (as not being discovered and proved by the true form of interpretation) I do not at all mean to bind myself. [IV, 31–2]

He regards such generalisations as 'interest payable from time to time until the principal be forthcoming' [IV, 31]. Some such provisional generalisations are also to be incorporated into the natural histories themselves, 'for they are useful, if not altogether true' [V, 136]. They are to be clearly labelled as working rules, not as scientific truths. As we have seen, Bacon was always scathing about those who mistook convenient working hypotheses (like the hypotheses of the astronomers) for certain principles.

Preliminary study of the material compiled in the natural histories enlarges the sum total of knowledge in one further way. Some configurations of circumstances (natural or contrived) lead to particularly striking observations about the phenomena involved. It is open to the investigator to contrive a new situation by analogy with an instructive instance, so as to make equally striking observations in other fields. The products of such use of analogy in natural philosophy are incorporated into the natural histories, and into the tabulation prior to the induction (see above, p. 130), and may incidentally have useful practical applications. Like the formulation of provisional generalisations the transferring of experiments from one field to another is done tentatively, without any guarantee of the results.

All provisional generalisation and all transferring of rules from one field to another by analogy depends upon the systematic examination of the material collected in the natural histories. I therefore begin the present discussion of Bacon's treatment of analogy and generalisation in natural philosophy with a consideration of the form the natural histories take, and the ways in which they are composed and extended. In the remaining sections of the chapter I discuss the implications of Bacon's experimental histories, and his advice on deriving from them working generalisations and additional material to enlarge the particular histories which form the basis for the interpretation of nature.

Natural histories

A natural history is an uncritical record of observations of natural phenomena, which corresponds to the store in the memory of primitive sense-perceptions. This means that the observations it contains are to be recorded without embellishment, without bias, and without supporting citations from classical sources, as concisely and perspicuously as possible [IV, 254]:

> I would have the History of Celestial Bodies simple, and without any infusion of dogmas; all theoretical doctrine being as it were suspended: a history embracing only the phenomena themselves (now almost incorporated with the dogmas) pure and separate; a history in short, setting forth a simple narrative of the facts, just as if nothing had been settled by the arts of astronomy and astrology, and only experiments and observations had been accurately collected and described with perspicuity. [V, 510–11][1]

[1] For Bacon, as I have remarked before, there is no problem about the possibility of giving a plain description which uniquely characterises a situation or event.

The material of the natural history is not to be organised for coherence, since any overall organisation will prejudice the outcome of the investigation.[1] Only two preliminary groupings of material are allowed: the material may be arranged according to a series of questions, or *particular topics* devised by the investigator, which focus attention on particularly important aspects of the subject;[2] and where a series of closely related observations are made these may be tabulated together [V, 135].

In addition to observations based on the past experience of the investigator, a history is to include observations based on experiments specially designed to yield further information, and to fill out sparsely represented areas of investigation. Observations and conjectures derived from other sources (for example, from classical authors) may also be included, to increase the available data, but the source must be clearly stated, and the content cautiously handled [V, 136]. Although Bacon recognises that the evidence of other investigators may introduce occasional errors into a history (particularly where the grounds for a belief are not stated), he believed that these do not occur frequently enough to upset the investigation, and that they can be recognised as invalid and eliminated in the course of analysis.

In Bacon's account of the compilation of sound natural histories 'experiment' is contrasted with 'accident'; experience sought for, with experience taken as it comes [IV, 81]. The term 'experiment' covers all situations deliberately sought by the investigator as a source of observations. These may be natural or contrived occurrences. They may or may not be the outcome of a search for particular effects. Thus although Bacon's use of the term covers what we should call 'experiments' it also covers a wide range of occurrences which we would not include. Many of the configurations of circumstances which Bacon calls 'experiments' we should merely allocate to the investigator's 'ex-

[1] Wherever Bacon lays emphasis on unbiased description he insists that material is not ordered, but written down as it is discovered by the investigator. In contrast, for teaching purposes material may be ordered to make it credible and accessible. See below, chapter 9.

[2] Bacon's particular topics are in the tradition of the dialectical places used to examine the available evidence (see chapter 1). They are 'places of invention and inquiry appropriated to particular subjects and sciences' [IV, 424]. Bacon however particularly emphasises that this list of topics should not be fixed, but should vary with the subject of investigation, and the extent of the investigator's knowledge of it [IV, 422–7]: 'For questions are at our command, though facts are not' [V, 135].

perience'.[1] And there is a further difference between modern and Baconian usage of the term 'experiment'. As the term is used today the scientist who believes a particular scientific theory (for instance that sound is transmitted by the air) contrives a test situation – an experiment – in which one outcome will be produced if the theory holds, and another if it is false. To test whether sound is transmitted by air, a bell may be suspended inside a glass vessel, from which the air is then pumped. If the bell is heard to ring whilst air remains in the flask, but not once it is evacuated, the theory is supported (the example is due to Boyle). Bacon's 'experiments' are not generally designed to test the truth or falsity of a scientific theory. Any observation illustrating some aspect of the topic under consideration which may prove useful in the future for *deriving* a theory is an experiment. Any observation of immediate practical benefit is an experiment. The former is an experiment of *light*, the latter of *fruit* [IV, 95, 17; V, 501]. The great merit of experiments of light, according to Bacon, is that they are random observations, to no particular end, which therefore further investigation without in any way biasing it:

> For the mechanic, not troubling himself with the investigation of truth, confines his attention to those things which bear upon his particular work, and will not either raise his mind or stretch out his hand for anything else. But then only will there be good ground of hope for the further advance of knowledge, when there shall be received and gathered together into natural history a variety of experiments, which are of no use in themselves, but simply serve to discover causes and axioms; which I call '*Experimenta lucifera*,' experiments of *light*, to distinguish them from those which I call '*fructifera*,' experiments of *fruit*. [IV, 95]

[1] In renaissance terminology the terms 'experiment' and 'experience' are used almost interchangeably. See the extremely interesting discussions of C. B. Schmitt. Schmitt argues that in the writings of Zabarella, and in the early works of Galileo, 'experiment' is used as a synonym for 'experience', and that Galileo's recognition of the need for a separate term ('periculum') to describe configurations of circumstances for testing a hypothesis is significant for the development of modern science. Elsewhere he shows the extremely theoretical nature of the 'experiments' of Aristotelian natural philosophers. It is not necessary to *perform* an experiment, only to suggest a situation which illustrates the point in question, and conjecture that the outcome will confirm that point. See Schmitt, 'Experience and experiment: a comparison of Zabarella's view with Galileo's in *De Motu*', *Studies in the Renaissance*, 16 (1969), pp. 80–138; 'Experimental evidence for and against a void: the sixteenth-century arguments', *Isis*, 58 (1967), pp. 352–66; P. Reif, 'The textbook tradition in natural philosophy, 1600–1650', *Journal of the History of Ideas*, 30 (1969), pp. 17–32, 30–1.

The experiment 'that locusts or crabs, which were before the colour of mud, turn red when baked' [IV, 258] is of no particular use in cookery (is not an experiment of fruit), but like the experiment that baked bricks turn from brown to red, it contributes to the investigation of the nature of 'redness' [IV, 258] (is an experiment of light for 'redness'):

> In like manner the fact that meat is sooner salted in winter than in summer, is not only important for the cook that he may know how to regulate the pickling, but is likewise a good instance for showing the nature and impression of cold. [IV, 258]

In insisting on more, and more varied natural histories, Bacon emphasises his preference for simple observations on the theme of the history, recorded virtually indiscriminately, rather than for instances which are immediately useful.

The basic collection of observations forms the *Primary History*. As far as one can gather the primary history is to be an enormous body of material presented as a continuous narrative, with general headings giving rough groupings of the contents (into 'concrete' [V, 135], or compound topics). The primary history falls into three parts: the history of the works of nature (the history of generations); the history of aberrations in nature (the history of pretergenerations); and the history of man's manipulations of nature (the history of arts) [IV, 294; V, 506]. Bacon regards it as particularly important that the last of these should not be treated entirely separately from the former two:

> And I am the more induced to set down the History of the Arts as a species of Natural History, because an opinion has long been prevalent, that art is something different from nature, and things artificial different from things natural...Whereas men ought on the contrary to be surely persuaded of this; that the artificial does not differ from the natural in form or essence, but only in the efficient; in that man has no power over nature except that of motion; he can put natural bodies together, and he can separate them; and therefore that wherever the case admits of the uniting or disuniting of natural bodies, by joining (as they say) actives with passives, man can do everything; where the case does not admit this, he can do nothing. Nor matters it, provided things are put in the way to produce an effect, whether it be done by human means or otherwise. [IV, 294–5][1]

[1] In traditional classifications of knowledge the arts appear in a category of their own within practical knowledge. See Reisch's classification above, chapter 4, p. 103.

The histories of generations and pretergenerations are taken together and subdivided into the history of celestial bodies, the history of meteors and the regions of the air (comprising histories for comets, meteors, winds, etc.), and the history of the 'exquisite[1] collections of matter', or history of species in nature. The history of arts is subdivided into history drawn from the mechanical arts, history drawn from the practical part of the liberal arts, and history drawn from areas of practical knowledge which have not as yet been fully systematised [IV, 257]. Histories for agriculture and navigation would fall under the first of these divisions, histories of medicine and music under the second, histories of baking, dyeing and tanning under the third [IV, 268–70].[2] The clearest idea we have of the scope of the primary history is the *Catalogue of Particular Histories by Titles* at the end of the *Parasceve* [IV, 265–70]. In this the histories are arranged according to the above divisions in a continuous list. It is noticeable that the titles at the end of some sections could equally fall at the beginning of the next section; the subdivisions are not intended to be hard and fast barriers between fields, but merely convenient subheadings.[3]

The histories of simple natures (dense, rare, etc.) are much narrower and more selective than any of the histories listed at the end of the *Parasceve*. They are histories of 'abstract natures' as opposed to 'concretes' [V, 135], and fall into a category of their own [IV, 29; V, 135]. Histories of simple natures are extracted out of the primary history [III, 552], and are sufficiently specialised by comparison with the primary history to be considered as 'a middle term between history and philosophy' [V, 510]. Such histories are to be compiled by experienced investigators like Bacon himself, who have acquired a good eye for the relevant and fundamental in nature, in the course of their observations and experiments [IV, 262].

The primary history is intended first and foremost to provide the initial stage in the interpretation of nature by means of the inductive method. The history of each simple nature is compiled

[1] 'Exquisite' or 'exquisitus' here means 'carefully sought out'.

[2] If we compare this once again with Reisch's classification we see that Bacon groups all practical activities together as intimately linked with the study of natural philosophy as a whole, rather than separating them out as independent activities (the crafts and domestic skills being inferior to theoretical investigation). See chapter 4, p. 103.

[3] This account of natural history is based on [IV, 251–70, 293–9; V, 208–10, 505–10].

in the light of the comprehensive general history [III, 552]. The contents of each such history are tabulated, and the form of each simple nature is discovered by applying the procedure of inductive elimination. The forms of all compound bodies are then presumably derived from the forms of simple natures in conjunction with the primary history; forms of compound bodies are built out of simple natures in such a way as to be consistent with all the relevant observations in the primary history. At this stage Bacon expects that the observations of the primary history will be corrected in the light of the certain knowledge yielded by the forms of simple natures [IV, 105, 260].

But Bacon also acknowledges the possibility of extracting useful generalisations in the course of compiling the primary history. Some such rules are to be incorporated in the preliminary histories as a guide to further investigation, and as a summary of investigations to date. The histories are therefore simultaneously records of short-term discoveries and the basis for the long-term task of discovering the certain first principles of all natural knowledge.[1] The writings intended to rally scientists to the task of compiling the primary history (and the king to finance the enterprise) particularly stress the fruitfulness of histories in producing general rules which may be tolerably reliable as the basis for practical activities. Natural histories are assigned a particularly prominent rôle in the Great Instauration on the grounds that they are a rich source of early benefits. Although the compiling of histories occupies third place in the *Plan of the Instauration* [IV, 22] it is to be embarked upon even before parts one and two are perfected (the comprehensive description of the state of human knowledge, and the *Novum Organum* or tool of interpretation itself):

Having therefore in my Instauration placed the Natural History...in the third part of the work, I have thought it right to make some anticipation thereof, and to enter upon it at once. For although not a few things, and those among the most important, still remain to be completed in my Organum, yet my design is rather to advance the universal work of Instauration in many things, than to perfect it in a

[1] Any of Bacon's own fragmentary histories shows clearly this double character. Bacon could never resist interspersing the histories with conjectural explanations of the phenomena, despite his avowed intention to subordinate all examination of material to the task of preparing it for application of the inductive method. See for instance 'provisional rules concerning the duration of life and the form of death' in *Historia vitae et mortis* [V, 320–35]. Bacon himself recognises this shortcoming of his histories.

few...when a true and copious history of nature and the arts shall have been once collected and digested, and when it shall have been set forth and unfolded before men's eyes, then will there be good hope that those great wits I spoke of before...after they have obtained proper material and provision will raise much more solid structures; and that too though they prefer to walk in the old path, and not by way of my Organum, which in my estimation if not the only is at least the best course. It comes therefore to this; that my Organum, even if it were completed, would not without the Natural History much advance the Instauration of the Sciences, whereas the Natural History without the Organum would advance it not a little. [V, 133–4]

Latent process and latent configuration

Physics, for Bacon, 'lies in a middle term between Natural History and Metaphysic' [IV, 347]. It investigates the 'vague, variable, and respective' rules in nature, where metaphysic investigates the constant and immutable laws.[1] This makes physics essentially a secondary study to metaphysic, 'for Physic carries men in narrow and restrained ways, imitating the ordinary flexuous courses of Nature; but the ways of the wise are everywhere broad' [IV, 362; see also III, 243]. All the techniques described by Bacon for discovering order and regularity in nature prior to application of the perfected inductive method fall within the investigations of physics [III, 555–6]. They are techniques for collating, transferring and extending knowledge of cause and effect derived from unsystematic observation, rather than from knowledge of the essential nature of bodies. To begin with, the local regularities in nature must be extracted from natural history in a tentative, rule-of-thumb fashion; after the perfection and application of the inductive method such regularities will be seen to be consequences of the forms of simple natures.[2]

[1] M. Primack gives a clear characterisation of physics and metaphysic in Bacon's scheme, which he summarises as follows: 'Physics as conceived by Bacon is concerned with the ordinary course of nature, i.e., matter as it is organized. Metaphysics (as the study of the forms of simple natures) is concerned with determining the exceptionless laws of nature, the existence of which is suggested to Bacon by his conception of matter as distinct from the world order.' See Primack, 'Outline of a reinterpretation of Francis Bacon's philosophy', *Journal of the History of Philosophy*, 5 (1967), pp. 123–32. Compare [IV, 126]. See also K. Lasswitz, *Geschichte der Atomistik vom Mittelalter bis Newton* (Hildesheim, 1963), Vol. I, Book 2, chapter 5, section 4, pp. 413–36.

[2] It must however be said that in presenting this secondary study Bacon is inclined to see it as a *parallel* study, and gives little indication of how the eventual programme of subsuming physics under metaphysic is to be realised.

The structure of a compound body discovered by simple experimental procedures such as dissection and distillation is its *latent configuration*. Investigations of latent configuration precede the completion of the *Novum Organum*, and the application of this tool to discover the forms of all simple natures. Some knowledge of the latent configuration of a body may enable one to perform simple operations upon it, following the course of nature, even though its form remains unknown. The following are some of the types of inquiry which Bacon believed would yield knowledge of latent configuration:

> What amount of spirit there is in every body, what of tangible essence; and of the spirit, whether it be copious and turgid, or meagre and scarce; whether it be fine or coarse, akin to air or to fire, brisk or sluggish, weak or strong, progressive or retrograde, interrupted or continuous, agreeing with external and surrounding objects or disagreeing, &c. In like manner we must inquire into the tangible essence (which admits of no fewer differences than the spirit), into its coats, its fibres, its kinds of texture. Moreover the disposition of the spirit throughout the corporeal frame, with its pores, passages, veins and cells, and the rudiments or first essays of the organised body, fall under the same investigation. [IV, 125]

Latent configuration is concerned with the interrelations between the gross parts of bodies, rather than with the concomitance of properties of the most fundamental parts. It is, in fact, the anatomy of the body [IV, 125], but an anatomy which digs deeper into the internal structure than, in Bacon's view, medical anatomy.

Without knowledge of latent configuration it will not be possible, according to Bacon, to effect any transformation of a body, since knowledge of forms must be combined with an understanding of the structure of the gross parts of the body to be transformed [IV, 124, 242]. Thus although in theory latent configuration plays a rôle secondary to that of forms and metaphysic, in practice it is an essential part of that knowledge which will empower man to transform natural bodies.

In a similar way, *latent process* is the natural, gradual alteration of bodies, which is made apparent by careful scrutiny of processes of development and growth in nature. Latent process may be imitated and adapted to produce unambitious transformations of bodies [IV, 122]. Where the study of forms will theoretically enable the investigator to generate gold (say) by the imposition of its essential properties (yellowness, malleability, and so forth), study of latent process will account for the way in which gold and other metals, and precious stones grow from the juices in which (according to Bacon [II, 374–5]) they originate. As part of this study the investigator is to record the losses and gains in substance, expansions and contractions in bulk, and so on, which accompany the development [IV, 124]. He will be able himself to initiate simple natural changes by imitating the stages of natural development.

As in the case of latent configuration Bacon apparently sees latent process as a necessary part of the knowledge which will enable man to transform bodies:

> Not only in the generation or transformation of bodies...but also in all other alterations and motions it should in like manner be inquired what goes before, what comes after; what is quicker, what more tardy; what produces, what governs motion; and like points...For seeing that every natural action depends on things infinitely small, or at least too small to strike the sense, no one can hope to govern or change nature until he has duly comprehended and observed them. [IV, 124]

Despite the fact that the superior study of the metaphysics of forms is supposed to give man complete control over nature, knowledge of latent process and latent configuration is an essential part of the application of forms in order to transform bodies. This is a fundamental weakness in Bacon's programme for the growth of human power over nature, and one which remains completely unresolved. Because Bacon fails to work out the relation between latent process and configuration, with their spirits, expansions and contractions, and the metaphysics of forms with its vocabulary of fundamental properties, the gap between operative and contemplative knowledge remains as broad in his scheme of knowledge as in any of the more conventional theories which he criticises.

Experientia literata

To a very large extent the discovery of latent process and configuration, and of explanations in terms of efficient and material causes, which together make up physics, depends upon collecting the right sorts of evidence, and devising the right sorts of experiment. Observations randomly collected together will only accidentally yield any useful knowledge; experience carefully planned and organised, whilst not immediately a reliable source of fundamental principles, will yield valuable provisional rules:[1]

> When a man tries all kinds of experiments without order or method, this is but groping in the dark; but when he uses some direction and order in experimenting, it is as if he were led by the hand; and this is what I mean by Learned Experience. For the light itself, which was the third way, is to be sought from the Interpretation of Nature, or the New Organon. [IV, 413][2]

Learned experience or *experientia literata* is the material of the natural history organised in such a way as to suggest to a perceptive mind the possibilities for enlarging knowledge by applying techniques successful in one field in similar fields, or by applying experiments successful on one type of material to similar materials.[3]

Experientia literata argues by analogy from situations in which a specified procedure has had a desirable outcome to new situations in which it may prove equally fruitful. Such a procedure does not legitimately rank as philosophy, but is rather an experimental skill. It is a 'sagacity', a 'hunting by scent' [IV, 413, 421]; a *practical* not a theoretical activity. New knowledge is discovered by ingenious adaptation of existing knowledge, rather than by formal inference from fundamental principles.[4] Imagi-

[1] It should be remembered that Bacon's objection to 'puerile induction', or the premature establishing of generalisations on the basis of too narrow and biased a selection of instances, was directed primarily at dialectical or rhetorical induction, where instances are adduced in support of a conjecture, and are supposed to clinch the question. Bacon's own use of inductive generalisation is supposed to be free from this fault, because the instances precede the generalisation.

[2] In this passage Bacon uses 'methodus' and 'ordo' interchangeably.

[3] Bacon uses *experientia literata* both to mean the organised and expanded histories [IV, 96], and the techniques used to produce them [IV, 413].

[4] Bacon also calls *experientia literata* the 'hunt of Pan', and gives a vivid characterisation of it in the fable 'Of the Universe, according to the Fable of Pan': 'As for the tale that the discovery of Ceres was reserved for this god [Pan], and that while he was hunting, and denied to the rest of the gods, though diligently and specially engaged in seeking her, it contains a very true and wise admonition; which is, not to look for the invention of things useful for life and civilisation from abstract philosophies, which are as it were the greater gods, even though they devote all their strength to the purpose; but only from Pan, that is from sagacious experience and the universal knowledge of nature; which oftentimes, by a kind of chance, and while engaged as it were in hunting, stumbles upon such discoveries. For the most useful inventions are due to experience, and have come to men like windfalls' [IV, 326].

native use is made of past observations of nature to extend the body of data based on experience and experiment. *Experientia literata* is therefore a creative rather than an intellectual activity, and an essential intermediary between the 'works' in which the true interpretation of nature originates, and the speculative forms to which these works give rise:[1]

> There is also another [practical branch of science] which is not altogether operative, yet does not properly reach to philosophy. For all inventions of works which are known to men have either come by chance and so been handed down from one to another, or they have been purposely sought for. But those which have been found by intentional experiment have been either worked out by the light of causes and axioms, or detected by extending or transferring or putting together former inventions; which is a matter of ingenuity and sagacity rather than philosophy. And this kind, which I noways despise, I will presently touch on by the way, when I come to treat of *learned experience* [*experientia literata*] among the parts of logic. [IV, 366]

Bacon divides the techniques for imaginative extension of past experimental knowledge into eight types: 'the Variation, or the Production, or the Translation, or the Inversion, or the Compulsion, or the Application, or the Conjunction, or finally the Chances, of experiment' [IV, 413].

The first way of extending experimental knowledge is by varying the materials of the experiment. Paper has traditionally been made from linen; one might vary the experiment, and see if paper could successfully be made from 'wools, cotton, and skins' [IV, 413]. One might also vary the operator (the efficient cause) in the experiment, rather than what is operated on (the material cause): amber attracts straws when rubbed with a cloth; one might see if it attracts the straws if it is heated rather than rubbed [IV, 414]. On the other hand, one might vary the

[1] In Bacon's view, creative processes are the outcome of comparison and sifting of singular experiences (as opposed to general notions of any kind), which are stored in the memory. From his incidental comments it appears that such manipulations of individual impressions is carried out by (or in) the imagination. Formal reasoning manipulates general notions according to the fixed rules of logic. It was standard in sixteenth century faculty psychology to recognise reasoning and generalisation which deals only with individual instances and experiences rather than with general notions as the province of imagination or of distinct cogitative faculties associated with the imagination. It is worth noting that in the stoic theory of knowledge analogy, comparison and transference of impressions were also regarded as unexceptionable fundamental operations of the mind. See G. Watson, *The Stoic Theory of Knowledge* (Belfast, 1966), and C. Bailey, *Lucretius' De Rerum Natura* (Oxford, 1947), Introduction, p. 58.

proportions of the materials: 'A leaden ball of a pound weight dropped from a tower reaches the ground in (say) ten seconds: will a ball of two pounds weight...reach the ground in five seconds?' [IV, 414]

The second way of extending experimental knowledge is by prolonging or repeating the experiment (*productio experimenti*). This may be done by repeated application of the same operation: if the product of a distillation is itself distilled will the product of the second distillation be even purer and more condensed than the first? [IV, 415] Experimental knowledge may also be extended by reproducing artificially operations observed in nature, or by transferring an experiment from one field of knowledge to another. Grapes which hang together ripen fast; the ripening of apples may therefore be accelerated by storing them in heaps [IV, 416]. Embalming preserves dead bodies; could the technique enlarge medical knowledge by suggesting a similar method of preserving live bodies? [IV, 417]

A successful experiment may also suggest another designed to explore the opposite effect to that produced in the original. Bacon called this *inversio experimenti*. Heat is observed to spread upwards as the result of a simple experiment. Can this experiment be modified so as to demonstrate that cold spreads downwards? [IV, 418] This technique is particularly prominent in the 'table of absence' for heat in the *Novum Organum*. A convex burning glass increases the heat of the sun's rays; will a concave glass diminish it? Can burning glasses produce any warmth from the moon's rays? [IV, 131]

Experiments may also be constructed to test the limitations of the known properties of bodies (*compulsio experimenti*). A magnet attracts iron; can either the iron or the magnet be treated in such a way that this property is lost (say by burning the magnet, or by steeping it in acid)? [IV, 418] Or the known properties of bodies may be exploited to derive results in further experiments (*applicatio experimenti*). If equal volumes of wine and water are compared the water weighs more than the wine; therefore if the weight and volume of a sample of wine are known it is possible to discover whether the wine has been watered (it will be too heavy for its volume if this is the case) [IV, 419]. Or a series of experiments having similar effects may be conjoined (*copulatio experimenti*). Roses can be made to bloom late either by picking off early buds, or by exposing the roots early in the year; by applying both these techniques the roses may bloom still later [IV, 420].

The final way of extending experiments Bacon calls *sortes experimenti*, the chances of experiment. This simply involves trying experiments so outrageous and outlandish that noone has thought to try them before. Bacon is clearly determined to leave no stone unturned!

As these examples should make clear, Bacon groups together as *experientia literata* all manipulation of experience (or as he terms it, of experiments) which depends upon drawing analogies between one situation and another. He sets the ability to detect such possibilities for extension of experience apart from mere *mechanic* – practical activity for its own ends, or to implement the findings of physics [IV, 366]. In Bacon's scheme of knowledge the products of random practical activity are simply to be incorporated into the natural histories [IV, 365–6]. Practical application of the provisional rules of latent configuration and latent process is mechanic proper [IV, 366]. But the astute creation of new experiments by analogy with old ones, either for their practical usefulness, or as a source for further knowledge to further the interpretation of nature by means of the *Novum Organum* is in a class of its own.[1]

Nevertheless, this special *experientia literata* is still for Bacon a study secondary to the metaphysics of forms. *Experientia literata* moves only from works to works; the interpretation of nature, or *Novum Organum*, moves from works to principles, and from principles to further works, ascending and descending to enlarge beyond measure the range of human achievement [IV, 413]. In the final analysis *experientia literata*, like the prerogative instances it so closely resembles, is merely a stage on the way to forms, intended to ensure the close intercourse between practical achievement and theoretical advance:

But after this store of particulars has been set out duly and in order before our eyes, we are not to pass at once to the investigation and discovery of new particulars or works; or at any rate if we do so we must not stop there. For although I do not deny that when all the experi-

[1] *Experientia literata* shares close links with the prerogative instances (see above p. 124). Both use 'sagacitas', applied directly to the works of nature, to enlarge the investigator's experience in telling ways.

ments of all the arts shall have been collected and digested, and brought within one man's knowledge and judgment, the mere transferring of the experiments of one art to others may lead, by means of that experience which I term *literate* [per istam Experientiam quam vocamus Literatam], to the discovery of many new things of service to the life and state of man, yet it is no great matter that can be hoped from that; but from the new light of axioms, which having been educed from those particulars by a certain method and rule, shall in their turn point out the way again to new particulars, greater things may be looked for. For our road does not lie on a level, but ascends and descends; first ascending to axioms, then descending to works. [IV, 96]

This intermediate rôle of systematic experimenting is made particularly clear in the account given in the *New Atlantis* of the hierarchy of scientific investigators in the ideal community. All practical adaptation and extension of experience is ranked below the interpretation of nature:

'We have three that collect the experiments which are in all books. These we call Depredators.

'We have three that collect the experiments of all mechanical arts; and also of liberal sciences; and also of practices which are not brought into arts. These we call Mystery-men.

'We have three that try new experiments, such as themselves think good. These we call Pioners or Miners.

'We have three that draw the experiments of the former four into titles and tables, to give the better light for the drawing of observations and axioms out of them. These we call Compilers.

'We have three that bend themselves, looking into the experiments of their fellows, and cast about how to draw out of them things of use and practice for man's life, and knowledge as well for works as for plain demonstration of causes, means of natural divinations, and the easy and clear discovery of the virtues and parts of bodies [this is *experientia literata*]. These we call Dowry-men or Benefactors.

'Then after divers meetings and consults of our whole number, to consider the former labours and collections, we have three that take care, out of them, to direct new experiments, of a higher light, more penetrating into nature than the former [these provide prerogative instances]. These we call Lamps.

'We have three others that do execute the experiments so directed, and report them. These we call Inoculators.

'Lastly, we have three that raise the former discoveries by experiments into greater observations, axioms, and aphorisms. These we call Interpreters of Nature.' [III, 164–5]

Nowhere else in Bacon's writings is the gulf between his own aspirations and modern science so dramatically clear. In treating systematic experimenting as subsidiary to a formal logic which manipulates essential definitions (albeit supposed to be empirically grounded), Bacon dismisses as secondary and provisional much of what we now regard as science.

8

Analogy and generalisation in ethics and civics

Ethics and civics were traditionally applied or practical subjects.[1] They contained contingent generalisations providing useful practical precedents for personal and political behaviour. Within ethics and civics themselves there were to be found no certain rules, both subjects being dependent for first principles on the philosophical studies of theology and metaphysics.

Bacon's treatment at first sight appears in sharp contrast with the traditional view, since he maintains that the inductive method is to be applied in all three branches of philosophy [IV, 112]. This is in line with his programme for uniting theoretical and practical branches of knowledge in all fields (see chapter 4). However, not even the outlines of such an application are to be found in his works.

There seem to be two main reasons for this. In the first place, Bacon like Aristotle accepts the fact that human behaviour tends to be much more complex than the behaviour of other natural phenomena. More disturbing factors interfere with the basic patterns, and the variables are less easily determined. Civics in particular is 'conversant about a subject, which of all others is most immersed in matter, and with most difficulty reduced to axioms' [V, 32]. There is also a theological difficulty, since knowledge of the principles which determine human voluntary behaviour is forbidden to man.[2]

[1] In *Nicomachean Ethics* I, 3 Aristotle points out that different subjects admit study to differing levels of certainty, and that ethics and politics are not exact sciences.

[2] Although Bacon's account of the corporeal basis for memorising, imagining and reasoning take him far in the direction of a mechanistic account of mental operations (though not as explicitly as Telesius'), he accepts the traditional account of the immortality and incorporeality of the rational soul [e.g. IV, 396–7], which governs these operations. The form of the rational soul is essentially involved in all human activities which involve decisions. Since the rational soul is incorporeal its form is inaccessible to man: 'Although I

It is therefore not surprising that despite Bacon's confident claims for the universality of the inductive method, traditional techniques for deriving useful precepts by generalisation and argument by analogy dominate his entire treatment of ethics and civics.

In this chapter I start by considering civil history and its relation to natural history, since civil history is for Bacon the main source of precedents and generalisations in ethics and civics.

Civil histories

In his classification of knowledge Bacon divides history into natural and civil:

> Natural History treats of the deeds and works of nature; Civil History of those of men. [IV, 293]

But here the similarity between the two branches of history ends. In describing events in nature, plain objective description is, according to Bacon, perfectly feasible. In describing the deeds of men such objectivity is neither possible nor desirable. The complex activities of men, and their interrelations with other men individually and in communities, require interpretation and assessment to be comprehensible at all. Hence the criteria for a 'perfect' natural history and for a 'perfect' civil history are of necessity different.[1]

am of opinion that [knowledge of the rational soul] may be more really and soundly enquired, even in nature, than it hath been; yet I hold that in the end it must be bounded by religion, or else it will be subject to deceit and delusion; for...it is not possible that it should be (otherwise than by accident) subject to *the laws of heaven and earth*, which are *the subject of philosophy*; and therefore the true knowledge of the nature and state of the soul must come by the same inspiration that gave the substance' [III, 379]. Hence that part of ethics and civics which involves rational rather than reflex activities can never be reduced to first principles derived from nature. As for the knowledge of the form of good and evil, which would provide the principles of ethics, Bacon presents its inaccessibility as follows: 'As for the knowledge which induced the fall, it was, as was touched before, not the natural knowledge of creatures, but the moral knowledge of good and evil; wherein the supposition was, that God's commandments or prohibitions were not the originals of good and evil, but that they had other beginnings, which man aspired to know, to the end to make a total defection from God, and to depend wholly upon himself' [III, 296–7].

[1] Bacon's bracketing of natural and civil history has led several historiographers to attempt to equate Bacon's criteria for the contents of natural history with those for civil history. They have therefore attempted to argue that Bacon advocates the same standards of detachment and objectivity in both fields, and hence that Baconian civil history heralds the advent of objective and 'factual' history, following in the wake of the colourful and literary histories of the early chroniclers and the Italian historians. L. F. Dean and G. H. Nadel both have difficulty in reconciling Bacon's own writings in civil history with the

Bacon's basic requirement for *natural* history is that it shall be terse, factual, and non-interpretative:

First then, away with antiquities, and citations or testimonies of authors; also with disputes and controversies and differing opinions; everything in short which is philological. Never cite an author except in a matter of doubtful credit: never introduce a controversy unless in a matter of great moment. And for all that concerns ornaments of speech, similitudes, treasury of eloquence, and such like emptinesses, let it be utterly dismissed. Also let all those things which are admitted be themselves set down briefly and concisely, so that they may be nothing less than words...it is always to be remembered that this which we are now about is only a granary and storehouse of matters, not meant to be pleasant to stay or live in, but only to be entered as occasion requires, when anything is wanted for the work of the *Interpreter*, which follows.
[IV, 254–5]

Natural history should in fact be a catalogue of observations roughly grouped according to subject matter.

Plain factual records concerned with human anatomy, physiology and affections are included within *natural* history. The *Catalogue of particular histories* with which the *Parasceve* closes [IV, 265–70] includes the following:

41. History of the Figure and External Limbs of Man, his Stature, Frame, Countenance and Features; and of the variety of the same according to Races and Climates, or other smaller differences.
45. History of Humours in Man; Blood, Bile, Seed, &c.
50. History of Voluntary Motions; as of the Instruments of Articulation of Words; Motions of the Eyes, Tongue, Jaws, Hands, Fingers; of Swallowing, &c.
77. History of the Affections; as Anger, Love, Shame, &c.
78. History of the Intellectual Faculties; Reflexion, Imagination, Discourse, Memory, &c. [IV, 267–9]

Histories like these provide material for scientific investigation of logic, ethics and politics, and in theory allow the interpreta-

objective criteria which he (supposedly) sets up for history in general. See L. F. Dean, 'Sir Francis Bacon's theory of civil history-writing', *English Literary History*, 8 (1941), pp. 161–83, reprinted in Vickers ed., *Essential Articles for the study of Francis Bacon* (Connecticut, 1968), pp. 211–35; G. H. Nadel, 'History as psychology in Francis Bacon's theory of history', *History and Theory*, 5 (1966), pp. 275–87, reprinted in Vickers ed., op. cit., pp. 236–50.

I find it difficult to sympathise with those historiographers who look to Bacon for a completely unbiased and objective civil history, since I find it difficult to imagine what sort of narrative they have in mind.

tion of nature to embrace human activities as well as other natural phenomena:

For I form a history and tables of discovery for anger, fear, shame, and the like; for matters political; and again for the mental operations of memory, composition and division, judgment and the rest; not less than for heat and cold, or light, or vegetation, or the like. [IV, 112]

The task of natural history (including the natural histories relating to man) is to provide natural data in a form in which it can immediately be used for the inductive derivation of first principles. *Civil* history only indirectly provides such material. Much of the material for the 'history and tables of discovery for anger, fear, shame, and the like' [IV, 112] will be derived from civil history, since the individual investigator's experience is unlikely to be extensive enough to provide sufficient examples to form the basis for an induction in these fields. But civil history provides such material precisely because (like 'poesy' or literature) it gives sympathetic and considered accounts not merely of human actions, but of the motives behind those actions, the states of mind of the participants, and possible interactions between events and personalities. Such information is not to be found in public records, and must to a large extent depend upon the critical judgment and general feeling, for a period of the historian himself. Bacon sketches the task of civil history and the historian as follows:

I come next to *Civil History*, properly so called, whereof the dignity and authority are pre-eminent among human writings. For to its fidelity are entrusted the examples of our ancestors, the vicissitudes of things, the foundations of civil policy, and the name and reputation of men. But the difficulty is no less than the dignity. For to carry the mind in writing back into the past, and bring it into sympathy with antiquity; diligently to examine, freely and faithfully to report, and by the light of words to place as it were before the eyes, the revolutions of times, the characters of persons, the fluctuations of counsels, the courses and currents of actions, the bottoms of pretences, and the secrets of governments; is a task of great labour and judgment. [IV, 302]

But such well constructed civil histories are rare. Amongst the features of extant histories which detract from their general usefulness are lack of order, lack of careful interpretation of events, partisan interpretation, and too free a hand in reconstructing where evidence is scarce [IV, 302]. For Bacon the quality of the explanation is of greater importance than the quantity of documentary evidence adduced.[1]

Bacon divides civil history into three types: bare records of past events; documentary fragments; and histories which select and interpret the events narrated. These he calls Memorials, Antiquities and Perfect Histories, respectively [IV, 303]. The first two types Bacon passes over as of their nature fragmentary and incomplete, and hence as useless as a foundation either for provisional generalisations, or for the natural histories concerned with man, until their contents have been analysed and organised. Whilst respecting the activities of antiquarians who sift such material and salvage what is of importance [IV, 303–4], Bacon considers such work as secondary to the writings of 'perfect' or narrative histories. The antiquarian's findings are part of the body of material which the author of a perfect history assesses as the foundation for his narrative; the perfect history itself is an immediate source of generalisation and precedent useful for life.[2]

In addition to civil history 'properly so called' [IV, 302], which falls into the above divisions, the 'History of *Learning* and *the Arts*' [IV, 300] is also part of the body of knowledge known generally as civil history. This provides a narrative account of the progress and growth of learning, and the social and geographical conditions under which learning flourishes. Such a history provides the basis for future decisions about what should be taught to students, when it should be taught, and how.

Perfect history is of three broad sorts. It may present all the

[1] Bacon's views on civil history are in line with conventional contemporary attitudes. On contemporary historiography see E. Fueter, *Geschichte der neuern Historiographie* (Munich/Berlin, 1936); F. Gilbert, *Machiavelli and Guicciardini: Politics and History in sixteenth-century Florence* (Princeton, 1965), chapters 5 and 6; F. S. Fussner, *The Historical Revolution: English Historical Writing and Thought 1580–1640* (London, 1962). D. S. T. Clark gives a clear and concise summary of the development of attitudes towards history in the period; Clark, *Francis Bacon: The Study of History and the Science of Man* (unpublished Ph.D. diss., Cambridge, 1970), chapter II, 'Theories of Historical Study and Writing in the Sixteenth Century', pp. 18–83. For the conformity of the sixteenth-century view of history writing as an exercise in interpretation, and a fund of ethical and political precedents, with classical views on the purpose of the study of history compare the above accounts with B. L. Hallward, 'Cicero Historicus', *Cambridge Historical Journal*, 3 (1931), pp. 221–37. Unfortunately all the above authors see progress in historiography as a movement towards ever more objective and 'factual' narrative – towards a study of history for its 'own intrinsic value' (Clark, p. 49), and 'the pursuit of truth as an end in itself' (Hallward, p. 221), whatever that means.

[2] On the secondary rôle of the antiquarian in the period see Clark, op. cit., pp. 41–6.

major political events during a set portion of time ('History of Times', or 'Chronicles'). It may recount the private and public actions of a particularly important historical figure ('Lives'). Or it may select a particular historical enterprise ('as the Peloponnesian War, the Expedition of Cyrus, the Conspiracy of Catiline, and the like' [IV, 305]) and describe its progress and development ('Narrations or Relations'). Alternatively, civil history as a whole may be divided into *pure* and *mixed* history. Mixed or *ruminated* history uses the narrative material of pure or perfect history as its text, and intersperses it with a commentary, in which generalisations and precedents derived from the text are suggested:

> For some men have introduced a form of writing consisting of certain narratives not woven into a continuous history, but separate and selected according to the pleasure of the author; which he afterwards reviews, and as it were ruminates over, and takes occasion from them to make politic discourse and observation. Now this kind of Ruminated History I greatly approve, provided that the writer keep to it and profess it. [IV, 310]

It is likely that Bacon has in mind Machiavelli's *Discorsi...sopra la prima deca di Tito Livio...*(1531) as the type of ruminated history.

Bacon assesses the merits of the various types of civil history according to their value as a reliable source either for observations suitable for inclusion in the natural histories for man, or for social and political precepts and precedents. Since Chronicles are on the whole concerned with broad outlines of historical events, over a wide range of countries and persons, it is not a particularly fruitful source for detail and precepts. Furthermore, Chronicles are intended to exhibit man's grandeur, and the triumph of moral rectitude over adversity. Hence the chronicler is inclined to propose motives for actions which are morally laudable, but actually unlikely.[1] Narrations offer a great deal more detail, since the scope of the history is a good deal narrower than that of the Chronicle, and the historian probably has more access to documentary evidence about his subject. However, Narrations are often either commissioned, as an

[1] For an example of the sort of motives which the chroniclers suggested for actions see Clark's account of Polydore Vergil's and Hall's versions of events in the reign of Henry VII; Clark, op. cit., pp. 278–342. See also below, p. 157.

official account of some political action, or written as an attack on an adversary's conduct in politics or war. Hence Narrations tend to be too biased to provide a very satisfactory basis for precepts.

Lives, on the other hand, provide details both of the private motives and conduct of the individual concerned, and his public actions and their outcomes. The author of a Life, like the author of the Narration, is dealing with a narrowly circumscribed topic, and is likely to have available a great deal of documentary and literary material on which to base his account. Lives, therefore, provide reliable information both about the character, personality, and emotions of the subject (for inclusion in the natural histories of the related topics), and a clear view of the interrelations between motives, events and their outcomes (from which political precedents and generalisations can be drawn).

Lives fulfil particularly well Bacon's condition for a successful history that 'Above all things (for this is the ornament and life of Civil History), I wish events to be coupled with their causes' [IV, 300–1]. By this Bacon means that the temperamental, social and geographical details, as well as the details of policy, which contribute to a particular chain of events, should be clearly laid out so as to make clear the causal way in which the sequence of events occurred [IV, 301].[1] Where the actual efficient cause of an event or action is not clear, it is the historian's job to provide a possible cause consistent with his interpretation of the episode and the evidence of the contemporary account he is working with.[2]

[1] The stipulation that causes should be linked with effects in history does not make Bacon a more scientific historian than his contemporaries. Blundeville, for instance, uses this conventional 'place' as a guide to history writing: 'In hystories things woulde be disposed according to their owne proper nature, and therefore sith in every action there must needs be a dooer, or worker, the hystorie muste first make mention of hym, and then shewe the cause that mooved him to doe, to what intent and ende, in what place, and with what meanes and instruments.' T. Blundeville, *The True Order and Method of Wryting and Reading Histories* (London, 1574), fo. F1r. (This work is a reworking of Acontius and Patricius.)

[2] See F. J. Levy, *Tudor Historical Thought* (San Marino, 1967), chapter VII, 'Politic History', especially p. 258: 'That causes frequently had to be invented was inevitable. Sidney had seen that, and Bacon understood it as well. But the invention of causes meant that both the imaginative and the reasoning faculties had to be used in writing history; and this cast doubt upon the use of the ensuing history as raw material for politics. It was not the past that taught lessons, but the interpreter of the past.' Bacon is quite content to take as evidence the lessons, and postulated causes of the interpreter of the past, since he has access to most information regarding the action or event.

The 'History of Henry VII'

Bacon's *History of Henry VII* is a 'life' on the pattern described in the *De Augmentis* account of civil history. It follows that we should look at this work not as an exercise in objective presentation of documentary evidence, which it clearly is not,[1] but as an attempt to throw into relief the general precepts and precedents implicit in the description of the events and their outcomes, together with such additional material as may be used as the basis for future studies, both as part of the interpretation of nature, and to determine policy in comparable situations. This, I think, is precisely what the *History of Henry VII* sets out to do. Using Hall and Polydore Vergil as his sources for the historical detail (together with documents made available to him on parliamentary transactions) Bacon constructs an extremely plausible account of the ways in which Henry's dominant character traits influenced his political decisions, and the prevailing conditions which determined the outcomes of these decisions, indicating the factors which Henry failed to take into account and which interfered disastrously with his plans. In addition Bacon attempts to provide material on parliamentary rulings during Henry's reign which might prove useful to future legislators. Bacon's analysis of causes and motives is significantly different from those of his sources,[2] and perfectly consistent with his own

[1] See Clark, op. cit., p. 282: 'Certainly Bacon intended it to be an accurate account of the reign and John Selden, historian and antiquarian, believed it to be one of the few contemporary histories based on documentary evidence. But we know that according to the historical theory to which Bacon subscribed historical truth depended as much on the degree of experience and perception which the historian brought to his work as on the reliability of his material, the standard of his scholarship and the degree to which he availed himself of primary sources.'

[2] For detailed discussion of the differences in the accounts of the reign of Henry VII of Vergil, Hall and Bacon see Clark, op. cit., pp. 278–342. Clark shows in detail how Bacon amends his account to stress the importance of Henry's dominating passions of suspicion and avarice on the shaping of his policies, and on the political events of his reign. I give here a single example taken from Clark's account: 'When Bacon returned to the affairs of Brittany and to the outbreak of war, it was to emphasise more strongly the way in which English policy was directed by the king's desire for financial profit. Once again, for Vergil and Hall, both of whom stressed Henry's just indignation over the forced marriage of Ann of Brittany to Charles VIII, the declaration of war was a straightforward matter with no ulterior motive. Once again, too, Hall emphasised how patiently Henry had fought for peace and how just was his present intention "to restore agayne the Britishe nacion to their auncient libertie, and to expulse the Frenche nacion, which thrusted (sic) for the blood, death and destruccion of the poore Brytones, clerely out of that duchy and country." Both writers noted how genuinely distressed Henry was to hear of the unpreparedness of Maximilian, both admitted that by treating with the French he was open to charges of

beliefs (as outlined in the *De Augmentis*) about the interactions between private and public causes in political affairs. To a very large extent he is content to accept the chroniclers' accounts of the actual chain of events, and well-established popular beliefs about Henry's character. These he weaves into a narrative which provides a rich source of instruction in precisely those studies of the influence of personal psychology on decisions and political policy, the diverse and dispersed pressures to be taken into account in predicting the course of events in public affairs, and the discrepancy between the intention of the agent and the outcome of his actions, which Bacon in the *De Augmentis* regards as important untouched fields of political knowledge.

It is part of Bacon's artistry as a historian that he presents both detailed and general observations in these controversial fields as part of an entirely plausible account of Henry's life and reign. The generalisations are readily seen to be true, and the sequences of events are convincingly causal, so that the reader has the impression of arriving at the general precepts by his own assessment of the material:

> For though every wise history is pregnant (as it were) with political precepts and warnings, yet the writer himself should not play the midwife. [IV, 311]

cowardice and avarice but both decided that his motive for peace was the desire for an honourable settlement in the face of fresh domestic dangers.

'Bacon followed a quite different line. The war had little to do with the fortunes of Brittany which Henry had by now given up for lost. Instead, the Breton cause was merely used as an excuse for a show of hostility towards France which the king hoped could be turned to his profit: "though he shewed great forwardness for a war, not only to his Parliament and court, but to his privy counsel likewise (except the two bishops and a few more), yet nevertheless in his secret intentions he had no purpose to go through with any war upon France. But the truth was, that he did but traffic with that war, to make his return in money" [VI, 119]. Nor were Henry's peace moves inspired mainly by any other motive. Far from being "sore unquyeted and vexed...appalled, and made sorrowful" by Maximilian's failure to materialise, a likelihood known to Henry before he sailed, he found it a welcome excuse to treat for peace and even suppressed the information until he had landed in France. Thereafter, his actions were dictated partly by the fear of future troubles but mainly by the desire for the most lucrative settlement compatible with considerations of honour and reputation. The peace was unpopular with the people, gentry and nobility and even the king himself appeared to be embarrassed by it. So ended his only major intervention in European affairs. In Bacon's eyes, Henry had shown not only prudence, courage and diplomatic skill in his own cause but also his customary failure to provide for long-term eventualities and a growing readiness to let his 'fast-handedness' dominate his policy. The conflict with France was not a simple struggle between justice and injustice, right and wrong, but a product of amoral politics and to be understood in terms of that interplay between personality and circumstance which Bacon made the subject of his analysis of the whole reign' (pp. 319–20).

The very fact that Bacon's account of the reign of Henry VII was for so long accepted as authoritative[1] shows how well he judges the balance between fact and interpretation.[2]

Generalisation and analogy in ethics

For early renaissance historians, civil history was a source of moral instruction, either at the general level of demonstrating the triumph of virtue over vice, or by offering models of virtuous living for imitation.[3] Bacon points out that authors who have suggested that their readers should follow the example of virtuous characters from classical antiquity have omitted to explain how they should set about emulating them [V, 3–4]. He himself divides the study of ethics into two parts: the general study of ethical principles, and the detailed study of human nature which will enable the moral teacher to 'indoctrinate' others (or convince himself) [V, 5].

The fundamental principles of ethics (the true goal of human life, the nature of the good, and so on) are to be arrived at by way of *philosophia prima*[4] rather than by disputation [V, 6–10; III, 229–30]. For instance, it is, according to Bacon, a universal principle of *philosophia prima* that 'whatsoever is preservative of a greater Form is more powerful in action' [IV, 338]. This rule, he claims, is self-evidently true in physics. It therefore greatly facilitates investigation in the more complex field of ethics:

There is formed and imprinted in everything an appetite toward two natures of good; the one as everything is a total or substantive in itself, the other as it is a part or member of a greater body; whereof the latter

[1] The first historian to attack Bacon's account on grounds of its lack of 'factual content' and 'objectivity' was W. Busch. Busch objects strongly to Bacon's dependence on other chronicles, and his failure to consult all available primary material. See Busch, *England under the Tudors, vol. I: King Henry VII*, transl. Todd (London, 1895), pp. 416–23.

[2] In chapter 9 I show how Bacon is prepared to exploit literary genres to put across unconventional ideas in all fields in an acceptable form. It is likely that he has this in mind also in the *History of Henry VII*. Novel ideas about psychology and political practice are presented as if naturally arising out of a plain narrative of Henry's reign. For contemporary views on the way to read histories see Blundeville, op. cit.; J. Bodin, *Methodus ad Facilem Historiarum Cognitionem* (Parisiis, 1566), especially chapters 2 and 3.

[3] This view of history as 'philosophy teaching by example' was held by Plutarch and Livy – history furnishes particular illustrations of general moral precepts. Petrarch and Erasmus also shared this attitude towards history. See F. J. Levy, op. cit., chapters I and II; E. Garin, *L'umanesimo italiano* (Bari, 1965); M. P. Gilmore, *Humanists and Jurists: Six studies in the Renaissance* (Harvard, 1963), I and IV ('The Renaissance Conception of the lessons of History', 'Fides et Eruditio: Erasmus and the study of history'); P. Burke, 'A survey of the popularity of ancient historians, 1450–1700', *History and Theory*, 5 (1966), pp. 135–52.

[4] See above pp. 104–6.

is in degree the greater and the worthier, because it tends to the conservation of a more general form...

This being set down and strongly planted, judges and determines some of the most important controversies in moral philosophy. [V, 7–8]

On the basis of the generality of the principle it is possible to decide that the active life is preferable to the contemplative (because it benefits society, rather than the individual, and hence 'preserves the greater form') [V, 8]. Also, that pleasure (which benefits only oneself) is inferior to virtue (which benefits others) [V, 9], and that the philosopher must involve himself in society, and not withdraw from it [V, 10]. *Philosophia prima*, according to Bacon, resolves these long standing disputes once for all:

So if the moral philosophers that have spent such an infinite quantity of debate touching Good and the highest good, had cast their eye abroad upon nature and beheld the appetite that is in all things to receive and to give; the one motion affecting preservation and the other multiplication;...and again if they had observed the motion of congruity or situation of the parts in respect of the whole, evident in so many particulars; and lastly if they had considered the motion (familiar in attraction of things) to approach to that which is higher in the same kind; when by these observations so easy and concurring in natural philosophy, they should have found out this quaternion of good, in enjoying or fruition, effecting or operation, consenting or proportion, and approach or assumption; they would have saved and abridged much of their long and wandering discourses of pleasure, virtue, duty, and religion. [III, 229–30]

It is particularly clear from Bacon's account of the range of applications of the universal rule that 'whatsoever is preservative of a greater Form is more powerful in action' [IV, 338] how heavily he relies upon suggestive possibilities for transferring the applications of words:

Iron in particular sympathy moves to the loadstone, but yet, if it exceed a certain quantity it forsakes its affection to the loadstone, and like a good patriot moves to the earth, which is the region and country of its connaturals; so again, compact and massy bodies move to the earth, the great collection of dense bodies; and yet rather than suffer a divulsion in nature and create a vacuum, they will move upwards from the centre of the earth, forsaking their duty to the earth in regard to their duty to the world. Thus it is ever the case, that the conservation of the more general form controls and keeps in order the lesser appetites and inclinations. [V, 7]

Here the force of the physical rule in ethics is confirmed by the fact that phrases like 'in particular sympathy [sympathia particulari]', 'forsakes its affection [amores illos deserit]', 'forsaking their duty to the earth [cessabunt ab officio suo erga Terram, ut praestent officium suum Mundo ipsi debitum]', are (as far as Bacon is concerned) appropriate both to ethics and to the physical world.[1] A striking feature in Bacon's attempts to systematise the 'inexact' sciences according to the principles of natural science, is this trust in what to us appear to be the metaphorical possibilities of words. When I come to discuss Bacon's use of analogy and metaphor as literary devices it will be necessary to keep in mind the fact that in appealing to verbal similarities for the universal principles common to all sciences Bacon deliberately endorses analogy (and accidental play on words) as serious tools of investigation in all sciences.

The knowledge which enables the moral teacher to set men on the path to virtue, by persuading them to uphold in particular instances the general principles of ethics, is essentially *practical.* It belongs, according to Bacon, with such knowledge as 'the knowledge of the diversity of grounds and moulds' which enables men to apply the principles of growth from natural philosophy to the practical field of agriculture, and the knowledge of 'the diversity of complexions and constitutions' which underlies the application of principles in medicine [V, 23]. In applying the principles of moral philosophy it is necessary to know the basic human temperaments, the types of force which compel men to act, and the remedies which will turn men from vice to virtue.

Although in the *Novum Organum* Bacon allows it to be thought [IV, 112] that the inductive method is to be used to provide a solid foundation for subjects such as these, in his preliminary

[1] Traditional teleological explanation of course also exploited the broad definition of goodness as 'that at which all things aim' (Aristotle, *Nicomachean Ethics* I, 1). See Aquinas, *Summa Contra Gentiles* III, 2 and 3, especially: 'For that every agent acts for some end clearly follows from the fact that every agent tends definitely to something definite. Now that to which an agent tends definitely must needs be befitting to that agent, since the agent would not tend to it save because of some fittingness thereto. But that which is befitting to a thing is good for it. Therefore every agent acts for a good' (chapter 3, II, 7, in A. C. Pegis ed., *Basic Writings of Saint Thomas Aquinas* (New York, 1944)). Both Aristotle and Aquinas uphold the good of the state as superior to that of the individual as a 'greater' good, i.e. bigger, because made up of many individuals; *Nicomachean Ethics* I, 2 (1094b9); *Summa Contra Gentiles* III, 17 (Pegis, op. cit., II, 27). But Bacon is both bolder in the verbal similarities he exploits, and more optimistic in the amount of detailed correspondence between fields that he expects.

study of practical ethics in the *De Augmentis* the procedure is much more orthodox.[1] The teacher is to formulate general precepts to guide his practical instruction, drawing on his own experience, and on the wealth of material concerning human temperament which is to be found in literature and particularly in civil history. Such material is to be collated and arranged to reveal the regularities and patterns in human behaviour, but Bacon's account of this procedure does not suggest the rigour of the inductive method:

> Not however that I would have these characters presented in ethics (as we find them in history or poetry or even in common discourse), in the shape of complete individual portraits, but rather the several features and simple lineaments of which they are composed, and by the various combinations and arrangements of which all characters whatever are made up, showing how many, and of what nature these are, and how connected and subordinate one to another; that so we may have a scientific and accurate dissection of minds and characters, and the secret dispositions of particular men may be revealed; and that from the knowledge thereof better rules may be framed for the treatment of the mind. [V, 22]

Similarly, poets and historians are the best source of knowledge of the affections which drive men's actions:[2]

[1] As we saw in chapter 4, ethics was traditionally a practical study of this sort. That is, it is guided by local generalisation based on the experience of the teacher, rather than by the syllogistic consequences of certain first principles (although these principles may be invoked in the course of deciding on a particular course of action). Aristotle considered that ethics was a suitable study only for the mature and experienced. He characterised 'sagacity' or 'practical wisdom', the intellectual gift of the moral teacher, as neither philosophy (theoretical) nor merely an art (a skill). It is the ability to make sound practical judgments by rules of local order, without knowledge of first principles: 'The man who is without qualification good at deliberating is the man who is capable of aiming in accordance with calculation at the best for man of things attainable by action. Nor is practical wisdom concerned with universals only; it must also recognize particulars. That is why some who do not know, and especially those who have experience, are more practical than others who know; for if a man knew that light meats are digestible and wholesome, but did not know which sorts of meat are light, he would not produce health, but the man who knows that chicken is wholesome is more likely to produce health.' *Nicomachean Ethics* VI, 7; see VI, 5–8, (1140a25-1141a30).

Sagacity was, as we saw in chapter 7, the intellectual attribute of the investigator gifted in *experientia literata*. Both in physics and ethics Bacon rates highly the sort of ability which allows the investigator or teacher to extend his experience by provisional generalisation, or argument by analogy from one case to another. He thus places these studies with the traditional applied subjects (including medicine and agriculture; see Reisch's classification, chapter 4, p. 103). In practice such 'hunting by scent' is the closest that Bacon comes to tackling the problem of relating theoretical and practical knowledge.

[2] For the bracketing together of poetry and history as vivid narratives of human nature and actions in classical writing, see M. I. Finley, 'Myth, memory and history', *History and Theory*, 4 (1964–5), pp. 281–302.

But to speak the real truth, the poets and writers of history are the best doctors of this knowledge, which we may find painted forth with great life and dissected, how affections are kindled and excited, and how pacified and restrained, and how again contained from act and further degree; how they disclose themselves, though repressed and concealed; how they work; how they vary; how they are enwrapped one within another; how they fight and encounter one another. [V, 23]

Such 'dissections' play a rôle in ethics analogous to that of latent process and latent configuration in physics. They yield an anatomy of temperament, affections and states of mind. This knowledge enables the moral teacher to select his moment for attempting to transform an individual's state of mind and plan of action, by manipulating as far as is possible the natural forces at work. Like his discussion of latent process and configuration, and *experientia literata*, Bacon's account of corrective moral guidance is full of insight into fruitful sources of information, and important areas of study. But like those discussions, the discussion of practical ethics makes little advance on traditional rule-of-thumb methods, despite the grand claims for the interpretation of nature as universal tool for knowledge.

Generalisation and analogy in civics

Civics, for Bacon, is a study apart from ethics:

Moral philosophy propounds to itself to imbue and endow the mind with internal goodness; but civil knowledge requires only an external goodness, for that suffices for society. [V, 32]

Civics is the study of what men do, not what they ought to do. This amoral attitude towards civics Bacon explicitly associates with Machiavelli[1] [V, 17]. Civil history, instead of providing *exempla* or illustration of abstract ethical principles, becomes, within this view of civics, a source of evidence in support of rules for guidance to efficient political action.

Machiavelli's political theory invites comparison with Bacon's.[2]

[1] The classic study of the relationship between Bacon's civics and Machiavelli's is G. N. G. Orsini, *Bacone e Machiavelli* (Genoa, 1936). I have used this extensively. For the growth of the 'amoral' attitude towards civics and history see Clark, op. cit., chapter 2.

[2] The other candidate for comparison with Bacon is, of course, Guicciardini. However, Machiavelli's works seem to have been better known in England in the period, and more closely associated with the new movement of secular political theory. See F. Raab, *The English Face of Machiavelli* (London, 1964); M. Praz, *Machiavelli and the Elizabethans* (London, 1928); V. Luciani, *Francesco Guicciardini and his European Reputation* (New York,

Like Bacon, Machiavelli claims that he has a new *method* which will yield the true and universal principles of politics. Machiavelli sets out to establish such general rules as can be extracted from examination of historical examples, assuming the single fact that human nature has remained the same throughout history. In practice he stays close to the traditional method of using historical examples to support previously determined general principles, but unlike earlier authors, his generalisations are political, not moral, and his examples contribute to the clear formulation, narrowing and qualification of general precepts.[1] The precepts in which he is interested are those concerned with establishing and maintaining a state in such a way that it is stable, that it interferes as little as possible with its members, and operates efficiently. The only 'morality' of his work appears to be that any actions directed to this end are 'right'.[2] Machiavelli chooses his terms (such as the term 'fraud' to describe political cunning or skill in manoeuvring) to emphasise the discrepancy between what is *right* in politics because effective in upholding the state, and what is *right* because divinely ratified as an action in keeping with man's true ends. The *Prince* is an exercise in debating the question 'what a principate is, what the species are, how they are kept, and how they are lost'.[3]

1936). Machiavelli also seems a more appropriate source of comparison because he, like Bacon, was intent on discovering general, practical rules, whereas Guicciardini stressed the individuality and unrepeatability of political events. See Luciani, op. cit., pp. 6–7 and pp. 388–9. See also F. Gilbert, op. cit., Part II. V. Luciani, 'Bacon and Guicciardini', *Proceedings of the Modern Language Association*, 62 (1947), pp. 96–113, shows that Bacon had read Guicciardini's histories. Bacon cites both Machiavelli and Guicciardini, the former more frequently. See Luciani, 'Bacon and Machiavelli', *Italica*, 24 (1947), pp. 26–40. I have based my assessment of Machiavelli's political philosophy on the following sources: A. Renaudet, *Machiavel* (revised ed., Paris, 1956), especially parts I and IV; L. Olschki, *Machiavelli the Scientist* (California, 1945); F. Chabod, *Machiavelli and the Renaissance* (Harvard, 1958), chapter 3.

[1] Olschki and Chabod both appreciate the fact that Machiavelli's 'method' cannot legitimately be characterised simply as induction by enumeration of instances. See Olschki, op. cit., p. 44, and p. 47: 'His theoretical presuppositions and the inner logic of his doctrines led him to conceive of examples and imitation in a more scientific than didactic sense. His examples are paradigms of experience and starting points for his inductive generalizations. They disclose the internal analogies of the historical events he considered typical, significant or decisive.' See Chabod, op. cit., chapter 3. S. Anglo, *Machiavelli: A Dissection* (London, 1969) also notes that examples are heavily doctored to support a previously suggested generalisation, and upbraids Machiavelli for failing to fulfil the conditions for a 'scientific' induction in his use of historical example.

[2] See Renaudet, op. cit., pp. 122–3; p. 136; Luciani, *Francesco Guicciardini and his European Reputation*, p. 1.

[3] Letter to Vettori, cit. Olschki, op. cit., p. 22. In this context the criticism of Gentillet, whilst couched in rather extreme language, sheds interesting light on Machiavelli's 'method'. In the preface to his *Discours sur les moyens de bien gouverner...un royaume...contre Machiavel*

Political upheaval is a necessary constituent of Machiavelli's system. A new political order is created by overthrowing an old; once in power, the preoccupation of the ruler is to preserve the state in unstable equilibrium, in the face of forces acting always to overthrow the ruling power. Such concerns are largely absent from Bacon's treatment of politics, as they are from contemporary English political writing as a whole.[1] Bacon assumes a monarchy, and an established political structure, and explores instead the opportunities for individual aggrandizement within this system. Although he claims that this self-advancement is to be achieved by morally acceptable means, and with a view to furthering the good of the community, Bacon's approach is, if anything, more amoral than Machiavelli's. He does not even put forward the well-being and stability of the state as the end product of his efforts, although he apparently imagines that 'self-good' and the good of the community ideally coincide.[2]

Bacon divides civics into three parts: the art of conversation, the art of negociation, and the art of government. The art of conversation is the art of putting up a good appearance. This Bacon claims does not simply make an individual well thought of, but raises him to positions of eminence because he gives the impression of integrity and all virtues [V, 33]. Once again this is a practical study, and involves precepts based on the experience of the courtier of long standing, which will enable the novice to advance himself.[3]

([Genève], 1576), Gentillet argues that in politics it is not appropriate to move deductively from 'causes et Maximes' to 'la conoissance des effects et consequences', or from 'effects et consequences' to deduce 'les causes et Maximes', without also considering the 'circonstances, dependances, consequences, et antecedences' which 'sont le plus souvent toutes diverses et contraires, de maniere que combien que deux afairs seront semblables, il ne les faudra pas pourtant conduire et determiner par mesme reigle ou Maxime, à cause de la diversité des accessoires' (p. 2). Gentillet maintains that Machiavelli's 'method' amounts to using the 'resolutive' method from effects to causes, and 'des particularitez aux Maximes generales', and that Machiavelli is mistaken in thinking that one can apply this method rigorously in politics, because of the many contingent factors involved. Moreover, maxims arrived at in this way cannot be assumed to be universal – corrupt generalisations will be derived from a corrupt state of affairs.

[1] See Raab, op. cit.; Clark, op. cit., chapter II.

[2] See Orsini, op. cit., chapter 4. But as Orsini also makes clear, there remains considerable tension in Bacon's writings on politics and ethics between the good of the individual, and the good of the community, and the problems implicit in the assumption that the state is the mediator between these two are never resolved; see p. 122.

[3] Such practical studies are in the tradition of Elyot's *Boke named the Governour*, and Castiglione's *Courtier*. Bacon, however, considers the practical problems simply from the point of view of success in public affairs, with no concern for an underlying ethical code. He uses a comparison which reveals particularly clearly his pragmatic approach: 'To conclude, this behaviour is as the garment of the mind, and ought to have the conditions of

The art of negotiation or business is the study of the conduct of individuals within the political system. Those skilled in this art (again, men of experience and practical insight) are able to advise others on how to manage their affairs, and to plan their own public careers for their own best advantage. Once again, Bacon is concerned simply with general precepts for success, without moral commentary. At the end of his discussion of political expediency he adds an apologetic coda in which he insists that all self-advancement is to be carried out within a moral framework – the individual must have good intentions. But the only reason Bacon can supply for confining oneself to the *bonae artes* is that the pursuit of one's own ends at any cost will lead ultimately to divine retribution [V, 75–8].[1]

In the *De Augmentis* Bacon gives an extended example of the third branch of civics, the art of government – 'Example of a Summary Treatise touching the Extension of Empire' [V, 79]. This is a version of the essay 'Of the True Greatness of Kingdoms', and is a generalised treatment of a topic which Bacon discusses in detail in his incomplete treatise *Of the True Greatness of Britain* (1608) [VII, 47]. This work gives an idea of Bacon's political *theory*, as opposed to the low-level practical rules of the two other branches of civics. The treatise lays down the general conditions for the strength and stability of a state, in order to demonstrate the need for union on an equal basis between Scotland and England. The work is strongly reminiscent of Machiavelli's writings, in its technique of first stating a general precept, and then defining its detailed relevance to a specified situation, expanding and supporting this discussion with historical *exempla*.[2] As a whole (Bacon gives a complete plan,

a garment. For first, it ought to be made in fashion; secondly, it should not be too curious or costly; thirdly, it ought to be so framed, as to best set forth any virtue of the mind, and supply and hide any deformity; lastly, and above all, it ought not to be too strait, so as to confine the mind and interfere with its freedom in business and action' [V, 34]. Swift notes the moral slackness in such an approach when he parodies the comparison in *A Tale of a Tub*: 'To instance no more; Is not Religion a *Cloak*, Honesty a *Pair of Shoes*, worn out in the Dirt; Self-love a *Surtout*, Vanity a *Shirt*, and Conscience a *Pair of Breeches*, which, tho' a Cover for Lewdness as well as Nastiness, is easily slipt down for the Service of both' (*A Tale of a Tub*, ed. Davis (Oxford, 1939), p. 47).

[1] Bacon's reference to Machiavelli as an advocate of 'evil arts' at this point is the only instance of his adopting the conventional contemporary attitude towards the Italian's writings. Both passages from Machiavelli are taken out of context.

[2] Bacon also takes similar stands to Machiavelli on several of the issues. Orsini (op. cit.) details the indebtedness of the treatise to Machiavelli. See chapter 2, 'Bacone come Interprete e Apologista di Machiavelli', especially p. 50. Orsini points out Bacon's emphatic approval of Machiavelli where he cites him in political matters.

although the details of the work are not complete) *Of the True Greatness of Britain* shows how Bacon recognises the significance of what is only implicit in Machiavelli's writings: the possibility of collecting together precepts based on experience and example (in this case precepts giving the requirements for a powerful, impregnable state), and arranging them so as to deduce the particular policy to be advocated in particular circumstances.

Bacon's list of precepts giving the conditions for the true greatness (political powerfulness) of a state runs:

First, *That true greatness doth require a fit situation of the place or region.*
Secondly, *That true greatness consisteth essentially in population and breed of men.*
Thirdly, *That it consisteth also in the valour and military disposition of the people it breedeth: and in this, that they make profession of arms.*
Fourthly, *That it consisteth in this point, that every common subject by the poll be fit to make a soldier, and not only certain conditions or degrees of men.*
Fifthly, *That it consisteth in the temper of the government fit to keep subjects in heart and courage, and not to keep them in the condition of servile vassals.*
And sixthly, *That it consisteth in the commandment of the sea.*

[VII, 48–9]

He gives a corresponding list of precepts stating what things are *not* necessary to the true greatness of a state.

According to Bacon, precepts one, two and six above are automatically satisfied for England. Taking into account the prevailing atmosphere of internal political unrest following the union of England and Scotland under James I, Bacon argues that in order to fulfil conditions three, four and five, the Scots should be naturalised as British subjects (thus becoming free citizens of a united kingdom, and potential valiant soldiers), and that a pretext should be found for embarking on a 'just war' close at hand, to expand Britain's power, and unify her peoples. It will then necessarily follow that England will become a great and powerful kingdom.[1]

The point to emphasise in Bacon's use of interplay between generalisation on the strength of past instances, and particular present circumstances is the *flexibility* of his approach. Like Machiavelli, Bacon is more concerned to establish a policy in the

[1] The documentary evidence for believing that this is the policy Bacon intended is given in the preface to the treatise [VII, 39–44]. This work, I think, justifies Orsini's claim that Bacon was the first Machiavellian to understand the less melodramatic consequences of the Italian's political philosophy.

present instance, using the past as a guide, than to set up the general precept as a universal principle. This flexibility is a general feature of his treatment of subjects which are 'immersed in matter' – material and efficient causes in physics, political and legal regularity in civics.[1] In such fields precepts are to serve as guides, to be adapted to the case in question, and where necessary modified to fit the contingencies of the present case.

[1] I have not attempted to discuss Bacon's use of maxim and provisional generalisation as the basis for reform of the legal code. This is a specialised issue which I do not feel qualified to handle. However, it appears to me that the *Maxims of the Law* set out generalisations formed on the basis of past legal cases, and to be amended or adjusted in the light of future particular cases, in the same way as I have described for generalisation in physics and politics. H. Jaeger gives a sympathetic account of Bacon's theory of law, and recognises the *ad hoc* nature of the *Maxims*; see Jaeger, 'Introductions aux rapports de la pensée juridique et de l'histoire des idées en Angleterre, depuis la Réforme jusqu'au XVIII siècle', *Les Archives de Philosophie du droit*, 15 (1970), pp. 13–70. P. H. Kocher gives a clear account of the *Maxims*, but assigns their formation to the inductive method, and then finds them inconsistent with Bacon's requirements for certain first principles; see Kocher, 'Francis Bacon on the science of jurisprudence', *Journal of the History of Ideas*, 18 (1957), pp. 3–26, reprinted in Vickers ed., *Essential Articles for the study of Francis Bacon* (Connecticut, 1968), pp. 167–94. See also B. McCabe, 'Francis Bacon and the natural law tradition', *Natural Law Forum*, 9 (1964), pp. 111–21; R. L. Stone de Montpensier, 'Bacon as lawyer and jurist', *Archiv für Rechts- und Sozialphilosophie*, 54 (1968), pp. 449–83; E. de Mas, 'L'origine della norma e della sanzione giuridica nel pensiero di Francesco Bacone', *Rivista Internazionale di Filosofia del Diritto*, 37 (1960), pp. 143–9.

9

Methods of communication

At the beginning of the second chapter of Book 6 of the *De Augmentis* Bacon introduces his treatment of *Method of Discourse*, the central study in 'the art of Transmitting, or of producing and expressing to others those things which have been invented, judged, and laid up in the memory' [IV, 438–9] as follows:

> [Method of Discourse] has been commonly handled as a part of Logic; and it also finds a place in Rhetoric, under the name of *Disposition*. But the placing of it in the train of other arts has led to the passing over of many things relating to it which it is useful to know. I have therefore thought fit to make the doctrine concerning Method a substantive and principal doctrine, under the general name of *Wisdom of Transmission*.
> [IV, 448]

There should be set up, according to Bacon, a separate study of the varied conventions used in transmitting a body of material to others. The foundation for such a study he recognises to be the precepts concerning *disposition* discussed briefly as part of rhetoric teaching, and those concerning *method* discussed by authors like Melanchthon and Ramus in the dialectic handbook.

On the face of it there is nothing startling about this remark. As I pointed out in chapter 1, Agricola, a century earlier, had included in his revised dialectic handbook a discussion of *dispositio* which drew together the treatments of material relating to this topic by rhetoricians and dialecticians. And Ramus' dichotomous method was supposed to characterise *dispositio* in the *ars disserendi* as a whole, incorporating areas of discourse traditionally discussed separately under rhetoric and dialectic. Nevertheless, Bacon's remark *is* extraordinary, and in the context of his outlook as a whole indicates an attitude towards *dispositio* which differs conspicuously from that of Agricola or Ramus.

Both Agricola and Ramus drew together rhetorical and dialectical discussion of disposition in order to emphasise the fact that the scope of their revised dialectic is broad enough to embrace both argumentative and literary composition. Agricola insists that insofar as all discourse sets out to communicate with others, it sets out to instruct, and hence may be called 'teaching'. His extension of the field of application of dialectic relies to a large extent on the claim that literary and informal discourse are in effect *more* rigorously based than traditional dialecticians would allow, and that a suitably framed dialectic will provide rules of composition for all such discourse. Ramus adopts the same position, though perhaps he is less eloquent in his formulation of it.

Bacon concedes with them that all the precepts for discourse of conventional dialectic and rhetoric are on a par. But for him this means that the precepts of dialectic and rhetoric are to an equal degree concerned only with plausible and persuasive discourse and with the swaying of opinions. Whilst they are 'very properly applied to civil business and to those arts which rest in discourse and opinion' [IV, 17], they are both entirely unsuitable as tools for exploring or reconstructing truth in any area of natural knowledge. Where Agricola implies that humanist, literary studies are as academically respectable and systematic as scholastic logic (and that a discipline is not made rigorous merely by proliferation of terminology), Bacon holds that both traditional and 'humanist' dialectic apply only in the sphere of oratory and sophistry.

Bacon is bound to consider even the reformed dialectic and rhetoric as second-class studies, because he is so deeply preoccupied with *discovery* as the primary mode of human experience. For him it is self-evident that as studies merely of *presentation* or transmission of the known and accessible, dialectic and rhetoric are both social, conventional arts, and in the last resort, parasitic activities. In the *De Augmentis*, 'method of discourse' is clearly presented as an art of 'producing and expressing to others those things which have been invented, judged, and laid up in the memory'; i.e. as the procedure which follows the separate collection and ordering of experiences, and formation of general precepts and principles [IV, 439]. In Book 5 Bacon takes great care to distinguish the 'invention' which collects facts from nature (invention 'of arts and sciences'), and the 'invention' which selects material for presentation to an audience (invention of 'speech and arguments') [IV, 407]. The latter is the technique associated with disposition or method of discourse:

> Now the use and office of this invention is no other than out of the mass of knowledge which is collected and laid up in the mind to draw forth readily that which may be pertinent to the matter or question which is under consideration...So (as I have said) this kind of invention [using the places or topics] is not properly an invention, but a remembrance or suggestion with an application. [IV, 421–2]

Such 'invention' has no natural privilege for Bacon, as Ramus had claimed for it, because only the inductive method mirrors our 'natural' apprehension of reality. Likewise, no order of presentation subsequent to this invention is privileged, since the inductive method alone effects 'natural' judgment of the phenomena. Ramus' dichotomous method, the subject of so much controversy (which 'hath moved a controversy in our time' [III, 403]), is therefore firmly assigned by Bacon to the collection of techniques which for aesthetic or psychological reasons are particularly effective as vehicles for transmitting new or difficult knowledge to an audience (like Agricola, Bacon takes it for granted that all communication is undertaken in order to teach something to someone); that is, to the *wisdom of transmission.*

Bacon uses the general term 'methodus' to describe all such techniques of presentation. He includes as types of 'method', along with Ramus' 'one and only method', 'delivery of knowledge in *aphorisms*' [IV, 450], '*assertions with proofs*' and '*questions with determinations*' (the traditional scholastic methods of teaching) [IV, 451], and the three Galenic teaching methods, 'Analytic, Systatic, Diaeretic' [IV, 452] (see chapter 1). All such formal methods of laying out material are, according to Bacon, conventions which are more or less successful in 'getting across' material to the student. Cautiously used they are an aid to presentation; immoderately used they distort the material:

> [The method of *questions with determinations*], if it be immoderately followed, is as prejudicial to the advancement of learning, as it is detrimental to the fortunes and progress of an army to go about to besiege every little fort or hold. For if the field be kept, and the sum of the enterprise pursued, those smaller things will come in of themselves;

although it is true that to leave a great and fortified town in the rear would not be always safe. In like manner in the transmission of knowledge confutations should be refrained from; and only employed to remove strong preoccupations and prejudgments, and not to excite and provoke the lighter kind of doubts. [IV, 451]

It is up to the teacher to organise his material within the limits of the convention (in this case, the dialogue question and answer form, such as Digby used in his *De duplici methodo*) so as to present his topic in the most favourable light, as comprehensibly as possible.

But Bacon considers more than just the formal conventions used by various schools for teaching, amongst the 'methods' which aid delivery of knowledge to others. For instance:

Next comes another diversity of Method...This is regulated according to the informations and anticipations already infused and impressed on the minds of the learners concerning the knowledge which is to be delivered...For those whose conceits [concepts] are already seated in popular opinions, need but to dispute and prove; whereas those whose conceits are beyond popular opinions, have a double labour; first to make them understood, and then to prove them; so that they are obliged to have recourse to similitudes and metaphors to convey their meaning...For it is a rule in the art of transmission, that all knowledge which is not agreeable to anticipations or presuppositions must seek assistance from similitudes and comparisons. [IV, 452]

Similitude and metaphor can be used to 'get across' abstruse knowledge to an ill-educated audience, and in this case, allegorical writing and extended metaphor used to clinch a point are 'methods' in their own right. So is parable:

Now this method of teaching, used for illustration, was very much in use in the ancient times. For the inventions and conclusions of human reason...being then new and strange, the minds of men were hardly subtle enough to conceive them, unless they were brought nearer to the sense by this kind of resemblances and examples...And even now, and at all times, the force of parables is and has been excellent; because arguments cannot be made so perspicuous nor true examples so apt. [IV, 317]

Literary conventions, like the formal conventions of teaching, allow the teacher to fit his material to the capacities of his audience, even if this means insinuating the knowledge in the form of parable and metaphor, and then 'justifying' it by pointing to the close match between the allegory and the teacher's interpretation of it.

By rejecting the claims that any of the existing methods of presentation are in any way privileged or natural ways to teach, Bacon sets all teaching methods on a par as devices for securing assent. He takes Agricola's view that 'teaching' includes all presentation whatsoever which sets out to hand over some knowledge to somebody. And the conclusion he draws is that there is no distinction in kind between the formal conventions of teaching (such as the disputation, the question and answer dialogue, dichotomous display of principles) and literary conventions which are more overtly manipulatory. The parable and its interpretation, the definition and its successive dichotomising, both equally take advantage of the reader's expectations of a convention. Dichotomies lead the reader to expect two members to the division at each stage, so that he will overlook material passed over as not fitting the dichotomy, and accept the two members even where they 'force [the material] out of its natural shape' [IV, 448]. The reader of a 'mythography' or collection of parables with interpretations expects a close and convincing match between the parable and the interpretation of its supposed hidden content, so that he will accept unconventional scientific or ethical beliefs where these are presented as a close interpretation of acceptable details of the fable.

Bacon, in his own writings, was preoccupied with the problems of presenting to a popular audience the abstruse and unorthodox conclusions of his own scientific and ethical theories. Amongst the many works in which he tried to present these ideas in a favourable light we have examples of profuse use of metaphor and comparison, extended use of parable and interpretation, aphorism, proverbs, and the essay form. In the remaining chapters of this work I shall look at some of Bacon's writings as examples of his putting into practice his belief in the broad and varied possibilities for using literary and pedagogic conventions as 'methods' for putting across new and unfamiliar ideas to a popular audience.

Initiative and magistral methods

Traditional teaching methods, in Bacon's view, do not give the reader the opportunity to assess the material presented himself, but lull him into uncritical acceptance by their very form. All the efforts of teachers of all times have been expended on making their material *plausible* by its presentation. There is, Bacon claims, no method which allows the student to examine the evidence, rather than accept the results:

> For as knowledges have hitherto been delivered, there is a kind of contract of error between the deliverer and the receiver; for he who delivers knowledge desires to deliver it in such form as may be best believed, and not as may be most conveniently examined; and he who receives knowledge desires present satisfaction, without waiting for due inquiry; and so rather not to doubt, than not to err. [IV, 449]

What is needed is an *initiative*, as opposed to a *magistral* method of delivery. Initiative methods are those which display the stages by which the author's conclusions were reached, so that the reader may both check that he would have reached the same conclusions on the same evidence, and pursue the investigation further if he so chooses. Since the inductive method is the sole means of arriving at sound conclusions from natural evidence, it is itself the ideal method for such presentation. Every stage in the derivation of scientific principles from sense data is recorded in the induction, so that the record itself will enable a future investigator to check the conclusions and pursue them further. Where such a rigorous method of inquiry has not been used, however, it is still possible for the author to set out his original assumptions and the way in which he arrived at his conclusions.[1]

Bacon's original observation that *all* artificial organisation and embellishment is inappropriate to scientific records was en-

[1] There is for Bacon no problem about the plain and unambiguous description of the process through which conclusions have been reached. When he complains of 'the first distemper of learning, when men study words and not matter' [III, 284], it is the preoccupation with eloquence at the expense of content that he is objecting to. 'For men began to hunt more after words than matter; and more after the choiceness of the phrase, and the round and clean composition of the sentence, and the sweet falling of the clauses, and the varying and illustration of their works with tropes and figures, than after the weight of matter, worth of subject, soundness of argument, life of invention, or depth of judgment' [III, 283]. Bacon evidently believed that if the author concentrates on giving a clear account of his topic, without affected concern for fashionable expression, the words he chooses will be a perfect match for the events described.

dorsed whole-heartedly by the founder members of the Royal Society. It may justifiably be regarded as a direct influence behind the Royal Society's policy of blunt, chronological reporting of the society's own proceedings, and of the experiments conducted by its members.[1] Bacon's stringent requirements for the initiative method mean that all his own writings (with the possible exception of the fragmentary histories of life and death, winds, dense and rare, etc.) are themselves excluded from this category. Even the *Novum Organum*, the least openly magistral of his works, fails to set out in its entirety the procedure by which Bacon arrived at the views presented to the reader.

Magistral methods to a greater or less extent try to impose their conclusions on the reader. Such presentation may be appropriate where the author's aim is for the novice to learn the foundations of some subject by rote, or to give his pupil a preliminary overview of a subject, such as (it is assumed in the period) most elementary education provides.[2] It was for just such purposes that the Ramist, dichotomising method had proved particularly useful, as a glance at almost any late-sixteenth-century teaching manual shows. Bacon objects to the use of this method in teaching, and blames it for the decay in learning in his times. According to him it gives a false air of completeness to the material, which leads to complacency, and discourages the student from inquiring further into the subject:

He thought also, that knowledge is uttered to men, in a form as if every thing were finished; for it is reduced into arts and methods,[3] which in their divisions do seem to include all that may be. And how weakly soever the parts are filled, yet they carry the shew and reason of a total; and thereby the writings of some received authors go for the very art: whereas antiquity used to deliver the knowledge which the

[1] See for instance T. Sprat, *History of the Royal Society* (London, 1667), p. 113.

[2] Grammar is a subject which lends itself to such presentation. In the period grammar was a largely conventional subject comprising precepts for correct word order in sentences, and the conjugation of nouns and declension of verbs. It is interesting to note that grammar is Ramus' example of a subject to be presented dichotomously in his *Dialectica*.

[3] In this context Bacon appears to use the term 'methods' to refer exclusively to the Ramist dichotomising method [See also III, 247; IV, 450]. This usage is responsible for the curious passage in *De Augmentis*, Book 6, chapter 2: 'Next comes another diversity of Method [Methodi], of great consequence to science; which is the delivery of knowledge in *aphorisms*, or in *methods* [Methodice]. For it is specially to be noted, that it has become the fashion to make, out of a few axioms and observations upon any subject, a kind of complete and formal art, filling it up with some discourses, illustrating it with examples, and digesting it into method [et Methodo revinciendo]' [IV, 450]. Here Bacon makes the popular Ramist method one of the particular divisions of 'method of discourse'.

mind of man had gathered, in observations, aphorisms, or short and dispersed sentences, or small tractates of some parts that they had diligently meditated and laboured; which did invite men, both to ponder that which was invented, and to add and supply further. [III, 498]

For they set [the sciences] forth with such ambition and parade, and bring them into the view of the world so fashioned and masked, as if they were complete in all parts and finished. For if you look at the method of them and the divisions, they seem to embrace and comprise everything which can belong to the subject. And although these divisions are ill filled out and are but empty cases, still to the common mind they present the form and plan of a perfect science. [IV, 85]

It is interesting to note that Bacon believed it to be a characteristic feature of divisive presentation that it creates an illusion of complete coverage:

That which consists of many divisible parts is greater than that which consists of few parts and is more one...

The fallacy here is very palpable, even at first sight; for it is not the plurality of parts alone, but the majority of them, which make the total greater. But yet this Sophism often carries away the imagination; yea, and deceives the sense. For to the sight it appears a shorter distance on a dead level, where nothing intervenes to break the view, than when there are trees and buildings or some other mark to divide and measure the space...So likewise in amplifications, the effect is increased if the whole be divided into many parts and each be handled separately. And if this be done without order and promiscuously, it fills the imagination still more; for confusion gives an impression of multitude; inasmuch as things set forth and laid out in order, both appear more limited in themselves, and make it evident that nothing has been omitted; whereas things that are presented confusedly, are not only thought to be numerous in themselves, but leave room for suspicion that there are many more behind. [IV, 470]

Unordered division of material gives the effect of profusion; ordered division adds the illusion that the coverage is complete, and the limits of the investigation clearly set.

Unordered division also conveys the impression that there is further material to be investigated. This is the characteristic which Bacon particularly favours in the method of 'aphorisms'. In aphoristic presentation each piece of information is digested into a terse, general, clearly comprehensible sentence (or series of sentences). These are then listed without artificial ordering or linking material. In this way each aphorism stands on its own merits, and the progress of the discourse is the cumulative effect of the discrete aphorisms. The fragmentary presentation of aphorisms leaves 'room for suspicion that there are many more behind' [IV, 470], so that others are led 'to contribute and add something in their turn' [IV, 451]. Equally, since no *context* is provided for each observation compressed into aphoristic form, the reader is at liberty to test its applicability in a variety of fields, and to explore not only its immediate consequences, but all its possible ramifications.

It should be noted that the aphorisms of this method of presentation are not witty, anecdotal sayings (although the term is used in this sense by other authors in the period). They are clear summaries of an aspect of the topic under discussion. The fact that aphorisms are designed to be *quotable* does not effect their seriousness. It is simply a measure of their intentional breadth of application. Solomon's *Proverbs,* considered as compressed statements of serious ethical truths, are Bacon's model for the successful use of aphoristic method.[1]

In his own works Bacon uses aphorism and division for just these calculated effects. Where his aim is to give a preliminary sketch of a field, and a sample of clearly and concisely stated precepts to spur the reader on to similar enterprises of his own, he uses aphoristic presentation. The *Novum Organum* is announced as exhibiting 'The Art itself of interpreting Nature, and the truer Exercise of the Intellect; Not however in the form of a regular Treatise, but only a Summary digested into Aphorisms' [IV, 35]. The preface to the *Maxims of the Law* explains that:

Whereas I could have digested these rules into a certain method or order, which, I know, would have been more admired, as that which would have made every particular rule, through his coherence and

[1] For a brief background to the use of the term 'aphorism' in the sense in which I discuss it here see B. W. Vickers, *Francis Bacon and Renaissance Prose* (Cambridge, 1968), pp. 61–70.

It is not correct to regard as an example of aphoristic method any use of terse sentence form in Bacon's writings, as, for instance, Vickers does, op. cit., chapter 3. Indeed, the same condensed sentence may function in one context as the grounds for an exploratory discussion of an ethical or scientific point (and hence be part of an aphoristic method of presentation), and in another context may provide illustration, and rhetorical force, as an example supporting a general point in discussion. Bacon uses Solomon's proverbs both as an example of aphoristic method, and, individually, as rhetorical *exempla.* See below, chapter 11.

relation unto other rules, seem more cunning and more deep; yet I have avoided so to do, because this delivery of knowledge in distinct and disjoined aphorisms doth leave the wit of man more free to turn and toss, and to make use of that which is so delivered to more several purposes and applications. For we see all the ancient wisdom and science was wont to be delivered in that form; as may be seen by the parables of Solomon, and by the aphorisms of Hippocrates, and the moral verses of Theognis and Phocylides: but chiefly the precedent of the civil law, which hath taken the same course with their rules, did confirm me in my opinion. [VII, 321]

In keeping with his condemnation of Ramist method and epitomes as the scourge of learning, Bacon never uses rigid dichotomous division. However, where he does wish to give an air of greater completeness of coverage he uses the less restricting rhetorical division or partition. This breaks the subject up into parts which clarify the progress of a discussion for the reader. The *Advancement of Learning* is laid out with the help of such partitioning, and the partition is extended and clarified in the later *De Augmentis.*[1] Such partitioning has all the advantages in persuading the reader of the range and completeness of the treatment which Bacon noted in the passage on the misleading powers of division quoted above (p. 175), without the distorting effect of strict Ramist division which 'produces empty abridgments, and destroys the solid substance of knowledge' [IV, 449].[2]

[1] For details of the changes in Bacon partition between the *Advancement of Learning* and *De Augmentis* see Vickers, op. cit., pp. 204–9.

[2] J. Hoskins, *Directions for Speech and Style* (1599?) quotes the passage from the *Colours of Good and Evil* [IV, 470] on the effect on the reader's imagination of division of the subject matter, in his treatment of division (*Directions for Speech and Style*, ed. Hudson (Princeton, 1935), pp. 22–4). For a contemporary account of the distinction between partition (dividing into convenient parts) and division (dividing into natural genera and species) see Melanchthon, *De Rhetorica libri tres* (Coloniae, 1521).

10
Parable

As I pointed out in the last chapter, Bacon believed that parable and its interpretation provides one alternative method for presenting unorthodox views to a vulgar audience, qualifying in this context as a legitimate teaching method. In this chapter I shall illustrate from the *De Sapientia Veterum* the way in which in practice Bacon uses parabolical method as a means of instruction.

As a teaching method, parables have the advantage of representing intellectual ideas in a more readily accessible form:

> Parabolical Poesy is typical History, by which ideas that are objects of the intellect are represented in forms that are objects of the sense.
>
> [IV, 315]

This is an advantage where the matter to be communicated involves concepts which are profound, or which require a radical shift in ideas on the part of the reader, before they can be grasped. In such cases the parable and its accompanying interpretation are a strategy for retaining the reader's attention, and carrying him through the stages in a difficult argument. The parable does not, of course, explain, or even justify the argument. It does, however, give an illusion of incontrovertibility to the material by virtue of the contrived match between the fable or parable and the subject matter of the interpretation.[1] Bacon regards parable and its accompanying interpretation as one method of overcoming the difficulties in passing on to an

[1] The power of persuasion is also increased by the fact that it was generally believed in the period that parable had been used to conceal abstruse knowledge by the ancients (Bacon includes this 'infoldment' of knowledge in parable as one of his methods of disposition). If the interpretation matches the traditional fable it may well discover the truth hidden in the fable by the ancients.

audience ideas at odds with their deepest preconceptions about nature and morality.

Bacon wrote one complete work of interpretation of classical myths or parables, the *De Sapientia Veterum*. Under the guise of interpretations of a series of traditional myths he presents some of the most abstruse and unconventional of his philosophical, political and ethical views in a form palatable to a popular audience. Judging from the frequency with which this work was reprinted he succeeded in placing his views in a setting in which they could be read, if not understood, by a wide audience.[1] I shall consider in detail the three myths from the *De Sapientia Veterum* which also appear as examples of parabolical poesy in the second book of the *De Augmentis*. We may, I think, assume that Bacon regarded these three expositions as particularly successful, since he reproduces them (and reworks them) in the later work. It is also convenient to select these three since they provide an example from each of the three traditional types of myth interpretation – philosophical, political and ethical. These are the myths of *Pan*, *Perseus* and *Dionysus*, respectively.

Bacon's account of the myth of Pan is extremely close to that of Comes, and one may assert with confidence that Bacon

[1] See [VI, 607–9] for the editions. It is interesting that Hoole, in his *A new discovery of the old arte of teaching schoole* (London, 1660) gives the *De Sapientia Veterum* along with Comes' *Mythologiae* as books suitable for reading in the sixth form at school. Hoole is extremely conservative, and would probably have disapproved of many of the views presented by Bacon as interpretation of the myths, if presented with them directly.

For a general background to myth interpretation in the sixteenth century see C. W. Lemmi, *The Classical Deities in Bacon: A Study in Mythological Symbolism* (Baltimore, 1933), introduction, pp. 1–45. Lemmi shows in detail in the body of his work how earlier and contemporary myth interpretations provided Bacon with material for his own work. However, he himself points out that few of the earlier works 'have any pretensions to literary excellence; they are student's companions or encyclopaedias'; Bacon 'transformed into an original thing of beauty a common but by no means artistic type of book' (p. 1). See also B. E. Carman, *A Study of Natalis Comes' theory of mythology and its influence in England* (unpublished Ph.D. diss., London, 1966), chapter I, 3. The search for classical and contemporary sources for Bacon's myths and interpretations is a vast undertaking. I have used Lemmi's account to indicate the most likely sources. Lemmi concentrates on the close resemblance between passages in Bacon's work, and passages in Boccaccio, *De genealogiis deorum gentilium* (1472?); N. Comes, *Mythologiae, sive Explicationis Fabularum libri decem* (Venetiis, 1567); Lucretius, *De Rerum Natura*; and selected alchemical writings. I have asumed that his painstaking collation of passages is reliable, and that he has not overlooked any obvious sources. I have restricted my own investigation to Comes' *Mythologiae* (Venetiis, 1567 and Parisiis, 1583 eds.); Boccaccio, *De genealogiis deorum* (Basilea, 1532 ed.); Cartari, *Le Imagini con la spositione dei dei de gli antichi* (Venetia, 1556 and Venetia, 1580 eds.). B. E. Garner, 'Francis Bacon, Natalis Comes and the mythological tradition', *Journal of the Warburg and Courtauld Institutes*, 33 (1970), pp. 264–97, indicates some of the respects in which Bacon adapts Comes' myth interpretations.

borrowed the myth from Comes' version in the *Mythologiae*.[1] However, when it comes to the question of the interpretation of the myth the situation is more complicated. Bacon certainly does not follow Comes uncritically, as he did for the details of the myth itself, although there are undoubted signs of the influence of Comes and other mythographers in isolated details of interpretation. We therefore need to give careful consideration to the question of the degree of originality displayed by Bacon in his reading of the myth, since only if there are significant divergences in the interpretation can we claim that his myth interpretation amounts to anything more than a rather elegant variation on an old theme.

The most striking feature of Bacon's interpretation is its consistency. Having identified Pan as 'the universal frame of things, or Nature' [VI, 709] ('Universitatem Rerum, sive Naturam' [VI, 636]) Bacon interprets the myth in all its details with reference to the hidden workings of nature, and appropriate tactics for exploring her secrets. He presents the myth as a source of confirmation for speculations about those aspects of

[1] *Mythologiae*, V, 6. With the exception of three points of detail Bacon's entire account is to be found here (1567 ed., fo. 137v.-140v, 'De Pane'). The three missing points are to be found elsewhere in Comes: in III, 6, 'De Parcis', Comes identifies the Fates as sisters to Pan (fo. 64r: 'ex illa confusa et informi materia cum Pane pastorum Deo natas fuisse'). In the 1583 ed. of the *Mythologiae* this point is inserted into the myth 'De Pane': 'Pronapis poëta in suo protocosmo natum fuisse Pana cum tribus sororibus Parcis e Daemogorgone scribit' (p. 448). The reference is to Boccaccio, op. cit., Book I, 'De Demogorgone'. In IX, 15, 'De Mida', Comes records the contest between Pan and Apollo (1567 ed., fo. 284v). In III, 16, 'De Proserpina' he identifies Iambe as the daughter of Pan and Echo (fo. 77v-78r: 'Erat Iambe muliercula quaedam Meganirae ancilla, ut tradidit Philochorus, Panos et Echus filia, quae cum Deam maestam videret, ridiculas narratiunculas, et sales Iambico metro ad commovendam Deam ad risum, et ad sedandum dolorem, interponebat'). Bacon omits some of Comes' speculations about Pan's parentage, but Comes himself restricts himself to the three accounts favoured by Bacon, when he comes to interpret the myth (fo. 139v). Otherwise Bacon's only notable omission is Pan's ruddy complexion, since this detail is standard in earlier mythographers' accounts (Comes, 1567 ed., fo. 140r). This detail occurs in Boccaccio, op. cit., 1532 ed., I, 4, p. 5: 'Moreover, by the redness of his face, I think they meant the element of fire which is connected to air, and joined thus together some people connect them with Jupiter' (transl. Carman, op. cit., II, 52); Cartari, op. cit., 1556 ed., p. 29: 'La faccia porporea, rossa, ed infocata.'

In the enlarged version of the myth in the *De Augmentis* [IV, 318–27] Bacon adds the story of Pan drawing the moon down by a deception (Comes, 1567 ed., fo, 139r), and the detail that the Fates lived in an underground cave (Comes III, 6).

The reader may convince himself that the closeness of the match between Bacon's and Comes' versions of the myth is more than coincidental by comparing the versions of the myth of Pan put forward by Boccaccio (op. cit., Book 9) and Cartari (op. cit., 1556 ed., pp. 131–42) which differ both in order of presentation and in numerous details. For the texts of fifteen mythographers' accounts of the myth of Pan see Carman, op. cit., appendix G. On the relation between Comes' myth of Pan and earlier versions see Carman, II, 3. For a table of the closer parallels between Bacon's and Comes' 'Pan' see Carman, appendix K.

nature to which experience and experiment can never give access: the origin and fundamental structure of the universe; the causal relations between natural phenomena; the interactions of natural bodies; the sort of experimental investigations which will give man unlimited power over nature.

This is in marked contrast to Comes' account, and indeed, to the general practice of the mythographers. Myth interpretation was basically eclectic, and aimed at being encyclopaedic, collecting together all available sources which might cast light on the myth. Comes and Boccaccio contrast alternative interpretations of Pan's physical characteristics and insignia, drawn from a variety of sources, without suggesting that any one of these is more likely than the others. Moreover, as a general rule interpretations in which Pan's physical attributes are identified with geographical features of the universe predominate over those in which Pan's attributes are interpreted as symbolising natural processes.[1]

Where the traditional interpretations do suggest explanations of details of the myth in terms of invisible natural processes Bacon does not hesitate in adapting and using these interpretations.[2] But his adaptations in themselves betray a concern with

[1] See for example, Boccaccio, op. cit., I, 4 (1532 ed., pp. 5–7): 'It remains to be seen what they were able to conclude about the image of Pan in which I myself think that the ancients wished to describe the whole body of nature, both of things animate and inanimate, inasmuch as, by his horns stretching up to the sky they wished to signify heavenly bodies, which we perceive to be two-fold, the one by art, which we know about through the investigation of the stars, and the other by perception whereby we feel their influence upon us. Moreover, by the redness of his face, I think they meant the element of fire which is connected to air, and joined thus together some people connect them with Jupiter...Moreover, they describe him as wearing a spotted skin, indicating by it the marvellous beauty of the eighth sphere, illustrated by the brilliance of its countless stars, by which sphere man is covered as if by a cloak, thus everything with regard to the nature of things is covered...I think that the lower parts of his body were hairy and shaggy to designate the surface of the earth, of the mountains, and the rough surface of rocks, forests, and thickets. Others thought that his image signified the sun who is believed to be the father and controller of things. Among these was Macrobius (lib I: 22 Sat.). And by his horns they wish to indicate the new moon, by his purple face the aspect of the aether when it is red at dawn and dusk, by his thick beard, the rays of the sun itself descending down into the earth, by his spotted skin, the adornment of the heavens, by his staff or crook, the control and moderation of things, by the pipe the harmony of the sky ascertained from the movement of the sun, as stated above' (transl. Carman, op. cit., II, 52–3). See Carman, op. cit., Appendix G for parallel texts of the myth of Pan from a selection of mythographers.

[2] Lemmi documents Bacon's borrowings for the Pan myth, op. cit., pp. 61–74. He does not, however, distinguish borrowings for the details of the myth from borrowings for the interpretation. The important borrowings for the interpretation are: Pan's hair and beard symbolising the rays from the sun – Macrobius (p. 62); Pan's goat feet symbolising upward striving of bodies – Macrobius and Comes (p. 63); the Fates symbolising natural causes – Boccaccio (pp. 65–6); Pan's birth – Comes (pp. 67–8); Pan's fight with Cupid symbolising tension between harmony and discord in nature – Comes (p. 69).

projecting views about nature which we know from Bacon's scientific writings to be basic to his own scientific beliefs. I take three interesting examples.

Comes interprets the two alternative versions of Pan's birth as follows:

Again they say that Pan was the son of Mercury, because since Mercury is the divine power and will, as we said, which brings things to their generation, and Pan is all natural, unmixed bodies together, these are governed by divine will...But since Pan contains all bodies of nature, as his name signifies, he is said to be born out of all things which exist, or to be made up of all things.[1]

No further details concerning the 'natural unmixed bodies' which make up the universe are given. But if we turn to the introduction to Book 2 of the *Mythologiae* we find Comes arguing that despite the variety of theories about the basic matter of which nature is constituted, all philosophers have agreed that one divine mind is responsible for natural order. The burden of Comes' argument seems to be that the variety of nature does not contradict belief in a single God behind it.[2]

Bacon's interpretation of the birth of Pan in the *De Sapientia Veterum* runs as follows:

Pan, as the very word declares, represents the universal frame of things, or Nature. About his origin there are and can be but two opinions; for Nature is either the offspring of Mercury – that is of the Divine Word (an opinion which the Scriptures establish beyond question, and which was entertained by all the more divine philosophers); or else of the seeds of all things mixed and confused together. For they who derive all things from a single principle, either take that principle to be God, or if they hold it to be a material principle, assert it to be though actually one yet potentially many; so that all difference of opinion on

[1] 'Pana rursus dixerunt esse Mercurii filium, quia cum Mercurius sit vis divina ac voluntas, ut diximus, quae res ad ortum perducit, ac Pan naturalia simplicia corpora universa illa a divina voluntate gubernantur...At quoniam Pan universa corpora naturae continet, ut nomen significat, dicitur, ex universis quae existunt esse genitus, sive ex omnibus consistere' (Comes, 1583 ed., pp. 453–4), transl. Carman.

[2] For Comes' philosophical beliefs see Carman, op. cit., chapter II, 2. Carman argues that myths often coloured Bacon's philosophical thinking, but that no individual interpreter (certainly not Comes) influenced his philosophy strongly. See I, 120: 'For the narratives of the accounts he relied extensively on Comes' *Mythologia* for genealogical and historical information. In the interpretation of these naturalistic myths, however, it becomes apparent that Bacon is paying little attention to tradition, and is interpreting these fables according to the findings of his own scientific experimentation.' This bears out my own findings. Elsewhere, however, Carman claims more direct influence on Bacon's interpretations of myths, and consequently on his philosophy; see e.g. p. 132.

this point is reducible to one or other of these two heads, – the world is sprung either from Mercury, or from all the suitors. [VI, 709]

Whilst preserving Comes' emphasis on the primacy of the Divine Will (and there can be little doubt that Bacon is influenced here both by Comes' interpretation of the myth, and by his argument in Book II for a single divinity), Bacon gives independent emphasis to the light the myth casts on natural order itself, rather than on its divine origin. In the *De Augmentis* version of the myth this is further emphasised:

For some philosophers have set down the seeds of things as infinite in their substance; whence arose the doctrine of *Homaeomerae* [sic], which Anaxagoras either invented or brought into repute. Some with greater penetration and judgment thought that the variety of things would be sufficiently explained, if the seeds were supposed to be in substance the same, but to take various, though certain and definite, figures; accounting for the rest by the position and connexion of the seeds one with the other; from which opinion emanated the doctrine of Atoms invented by Leucippus, and sedulously followed by Democritus. Others, though they asserted one principle of things (as Thales, Water, Anaximenes, Air, Heraclitus, Fire), yet maintained that principle itself to be actually one, but potentially various and dispensable, as that which had latent within it the seeds of all things. But those who (like Plato and Aristotle) have represented Matter as entirely despoiled, shapeless, and indifferent to forms, have approached much nearer to the figure of the parable. For they have made Matter as a common harlot, and Forms as suitors. [IV, 319–20]

Here the various opinions of classical philosophers concerning the basis for natural order (drawn probably from Cicero's *De Natura Deorum*) are presented as matching the myth in varying degrees. The theory of a material substratum on which forms are impressed, which we know Bacon favoured on other grounds (see p. 113), is presented as the best match for the details of the fable, and hence, by implication, as the most likely explanation. One of the more abstruse of Bacon's philosophical beliefs is thus given an appealing presentation.

In Macrobius, Boccaccio and Comes, Pan's beard is said to signify the rays of the sun.[1] In the *De Sapientia Veterum* and the

[1] See Lemmi, op. cit., p. 62. Comes puts it as follows: 'alii cornua Pani tanquam Soli tribuentes barbam eius promissam Solis ipsius lumen esse dixerunt' (1584 ed., p. 455).

De Augmentis Bacon adapts this interpretation in the following way:

> The body of nature is elegantly and truly represented as covered with hair; in allusion to the rays of things. For rays are as the hairs or bristles of nature, nor is there anything which is not more or less radiant. This is seen most evidently in the faculty of sight, and no less in all magnetic virtue, and every effect which takes place at a distance. For whatever produces an effect at a distance may be truly said to emit rays. But Pan's hair is especially long in the beard; because the rays of celestial bodies, especially of the sun, operate and pierce from a greater distance than any other; so that not only the surface, but even the interior of the Earth for some distance, is changed, wrought, and filled with spirit by them. [IV, 322]

Once again, the argument of the interpretation is carried along by its close match with the mythical detail. The detail (that Pan's beard is 'especially long' [IV, 318]) is added by Bacon to the preliminary account of the myth, and then matched to a telling point of interpretation, emerging as a peculiarly appropriate insight on the interpreter's part.

Finally, Comes interpreted Pan's love for Echo as the universe's enjoyment of the music of the spheres.[1] Bacon's account runs:

> The world therefore can have no loves, nor any want (being content with itself), unless it be of *discourse*. Such is the nymph Echo, a thing not substantial but only a voice...But it is well devised that of all words and voices Echo alone should be chosen for the world's wife; for that is the true philosophy which echoes most faithfully the voices of the world itself, and is written as it were at the world's own dictation; being nothing else than the image and reflexion thereof, to which it adds nothing of its own, but only iterates and gives it back.
> [IV, 326–7]

As far as I am aware, this exploitation of the characteristics of echoes (rather than of the fable of the nymph Echo) as a symbol for the true philosophy, which is the reflection of nature itself, is original to Bacon. Bacon's powerful use of parallel exposition lends credence to philosophical views to which he attached particular importance.

[1] 'Fama est praeterea Echo fuisse a Pane amatam, quippe cum coelorum harmoniam Echo esse putarent, quae redundaret e ratione motuum' (1584 ed., p. 456).

Political and ethical myths

In the myths of Perseus and Dionysus Bacon is more selective in his use of Comes. Whilst all the details of Bacon's versions of these myths are to be found in Comes, many further details are omitted, and Bacon makes no attempt to use Comes' order and manner of presentation.

Bacon's account of the myth of Perseus occurs in the *Mythologiae* as part of the myth of Medusa.[1] Comes interprets this myth as a lesson in humility. Medusa was turned from the most beautiful of women into a monster from whom men averted their gaze, because of her pride and concupiscence. Perseus' decapitation of Medusa (who turned men to stone with her glance) with the help of Pallas, is interpreted as meaning that men must seek divine help in order to overcome the sin of voluptuousness. Bacon ignores these interpretations, and instead uses the myth as the basis for a detailed tactical discussion of the art of war.[2]

Bacon's interpretation falls into two parts. The first part elaborates the account in the myth of Perseus' quest for Medusa into a discussion of favourable conditions for the undertaking of a war. This part owes little, as far as I have been able to discover, to the accounts of earlier mythographers.[3] In the second part Bacon interprets Perseus' weapons (the gifts of Pallas), and his slaying of Medusa, in terms of the preparations necessary for conducting a successful war, and here Bacon's account is more traditional.

In the first part Bacon uses the obscurer details of the myth as simple cues for fairly elaborate discussion of the strategy of war. He does not dwell on the match of detail and interpretation, and he digresses into historical *exempla* (particularly in the *De Augmentis* version of the myth), to support his tactical points independently of the myth. The interpretation is more openly literary than, for instance, the interpretation of the Pan myth, and the theme is more carefully sustained than in traditional inter-

[1] Myth VII, 11. See also the myth of the Gorgons, myth VII, 12. All the details of Bacon's version occur in Comes' account, including the visit to the Graeae, the Gorgons' sisters, the fact that Medusa was sleeping when Perseus found her, but that nevertheless he used Pallas' mirror to direct his blow, and the birth of Pegasus from Medusa's blood.

[2] Lemmi gives Fulgentius' *Mythologicon* and Boccaccio as ultimate sources for this interpretation, op. cit. pp. 156–9.

[3] It does show the influence of Machiavelli. See Lemmi, op. cit., p. 159.

pretations. The following extract gives the characteristic tone of this first part, showing how the details of the myth are worked into a narrative account of precepts for the guidance of princes and generals in warfare:

The fable seems to have been composed with reference to the art and judicious conduct of war. And first, for the kind of war to be chosen, it sets forth (as from the advice of Pallas) three sound and weighty precepts to guide the deliberation.

The first is, not to take any great trouble for the subjugation of the neighbouring nations...And therefore Perseus, though he belonged to the east, did not decline a distant expedition to the uttermost parts of the west.[1]

The second is that there be a just and honourable cause of war: for this begets alacrity as well in the soldiers themselves, as in the people, from whom the supplies are to come...Now there is no cause of war more pious than the overthrow of a tyranny under which the people lies prostrate without spirit or vigour, as if turned to stone by the aspect of Medusa.

Thirdly, it is wisely added that whereas there are three Gorgons (by whom are represented wars), Perseus chose the one that was mortal, that is, he chose such a war as might be finished and carried through, and did not engage in the pursuit of vast or infinite projects.

[VI, 715–16]

In the expanded version of the interpretation in the *De Augmentis*, Bacon gives the strong impression of using the myth as a convenient setting for a discursive account of war as political strategy.

In the *De Sapientia Veterum*, the second part of the interpretation contains a cursory account of the professional gifts of the competent general – Mercury's winged shoes stand for swiftness of action; Pluto's helmet for secrecy; Pallas' mirror and shield for foresight. In the *De Augmentis* these are expanded and linked into a much fuller account of the tactics to be used in planning and conducting a campaign of war. The following extract gives some idea of the way in which Bacon uses a scattering of mythical details as the framework for a condensed account of such tactics:

[1] It is traditional to make Perseus of eastern origin, and to place the Gorgons in the west. I have not, however, found another writer who links these two geographical locations as part of the interpretation, as Bacon does.

And the helmet of Pluto (which used to render men invisible) is a manifest parable. For next to speed in war secrecy of counsels is of the greatest moment; of which indeed speed itself is a great part; for speed anticipates the disclosures of counsels. To the helmet of Pluto belongs also this: that there should be one commander in a war, with free instructions; for consultations held with many savour more of the crests of Mars than the helmet of Pluto. Variety of pretexts, ambiguous directions, rumours spread abroad, which either blind or avert men's eyes and involve the real designs in obscurity, refer to the same. So also diligent and suspicious precautions respecting despatches, ambassadors, deserters, and many like matters, are wreathed round the helmet of Pluto. But it is of no less importance to discover the counsels of the enemy than to conceal our own. To the helmet of Pluto therefore must be added the mirror of Pallas, whereby to discern the strength or weakness of the enemy, their secret partisans, their discords and factions, their movements and designs. But since there is so much of chance in war, that no great confidence can be placed either in discovering the designs of the enemy, or in concealing our own, or even in speed itself, we must take special care to be armed with the shield of Pallas, that is, of foresight, so as to leave as little as possible to fortune. To this belong the exploring of roads before a march, the careful fortification of the camp (which in modern warfare has fallen almost into disuse, whereas the camps of the Romans were like a fortified town, to fall back upon in case of defeat), a firm and well drawn up line of battle, not trusting too much to light troops, or even to cavalry; in a word, everything which relates to a sound and careful system of defensive war; for the shield of Pallas is generally of more avail in war, than the sword of Mars itself.

[IV, 330–1]

In the *De Sapientia Veterum*, and more explicitly in the *De Augmentis*, Bacon uses the myth of Perseus and Medusa as the basis for a detailed presentation of a particular view of warfare; the provisions which should be made and precautions which should be taken before and during battle, in order to ensure success in a war. To this end he phrases the account of the myth which precedes the interpretation, so as to highlight key points in his discussion,[1] whilst omitting any extraneous detail which

[1] In the *De Augmentis* Bacon emphasises two points in the preliminary account of the myth which in the *De Sapientia Veterum* occur for the first time in the interpretation itself. These are, that Perseus came from the east to kill Medusa in the west, and that Mercury's wings were attached to the feet rather than to the shoulders. This helps to reinforce the impression that the myth convincingly establishes in advance the precise points made in the interpretation which follows.

might blur the clear match of myth with interpretation.[1] This is equally the case in Bacon's version of the myth of Dionysus, which he interprets as a moral allegory for the workings of desire in men. Once again, the details which Bacon chooses to include in his account of the fable are all to be found in Comes, but Bacon is selective, and omits a great deal of further (and confusing) detail from Comes' account.[2] His terse catalogue of attributes of Dionysus is in striking contrast to Comes' rambling narrative, and is asserted with deceptive confidence (since Dionysus or Bacchus was the subject of an embarrassingly rich and contradictory mythological tradition). As in the case of the Perseus myth, Bacon selects one theme from amongst the traditional interpretations of the myth,[3] and presents a series of moral precepts in the form of an interpretation. On this he builds a strikingly smooth and unified development of the theme of his interpretation:[4]

The mother of all desire (though ever so hurtful) is nothing else than apparent good. For as the mother of virtue is real good, so the mother of desire is apparent good. One the lawful wife of Jupiter (in whose person the human soul is represented), the other his mistress; who nevertheless aspires, like Semele, to the honours of Juno. Now the conception of Desire is always in some unlawful wish, rashly granted before it has been understood and weighed; and as the passion warms, its mother (which is the nature and species of good), not able to endure the heat of it, is destroyed and perishes in the flame. Then the progress of Desire from its first conception is of this kind. It is both nursed and concealed in the human mind (which is its father); especially in the lower part of it, as in the thigh; where it causes such prickings, pains, and depressions, that the actions and resolutions of the mind labour and limp with it. And even when it has grown strong with indulgence and custom, and breaks forth into acts (as if it had now accomplished its time and were fairly born and delivered), yet at first it is brought up for

[1] Bacon omits discussion of both Perseus' and Medusa's histories prior to their encounter, although such accounts were standard (see Fulgentius (Lemmi, op. cit., p. 157), Comes and Boccaccio).

[2] In particular, Bacon includes the details that Dionysus subdued the whole world as far as India, that his chariot was drawn by tigers and accompanied by Cobali, Acratus and the Muses, which occur in Comes, but are not typical of other mythographers' accounts of the Dionysus myth.

[3] Comes links Dionysus with desire, but on the whole favours an interpretation of Dionysus as Wine, or wine-making and consuming.

[4] In the *De Augmentis* version of this myth Bacon adds a remark against heretics, and a warning that awareness and advice cannot master a man's ruling passion [IV, 335]. On the whole, however, he seems content with the earlier version.

a time by Proserpine; that is, it seeks hiding-places and keeps itself secret, and as it were underground; until throwing off all restraints of shame and fear, and growing bolder and bolder, it either assumes the mask of some virtue, or sets infamy itself at defiance. [IV, 333]

The care with which Bacon parallels the phrasing and development of his account of the myth in his treatment of the psychological processes of desire is in marked contrast to the cursory accounts of more traditional mythographers. Where they are content to indicate the sort of human failings and conflicts which a particular myth might be considered to represent, Bacon draws out the possibilities for describing in the same terms events and relationships in the myth, and processes and interactions in the mind.[1]

In assessing Bacon's excursions into mythography in the *De Sapientia Veterum* and the *De Augmentis*[2] we need to consider the philosophical myths separately from the ethical and political myths. As I have shown in detail for the myth of Pan, Bacon leans extremely heavily on Comes' *Mythologiae* for myths traditionally considered to refer to natural processes. In his interpretations, however, he discards the accounts of traditional mythographers, although he incorporates some appealing details from Comes and others. He presents views which we know he held on other grounds (as part of a consistent scientific outlook) in such a way as to emphasise the close match between just such views and the details of the traditional myth. The myths thus appear to provide independent confirmation for particular philosophical and scientific beliefs.

Moreover, where his views changed in the interval between two versions of the same myth and interpretation, he contrives ways of attaching his changed views to the same mythical detail which previously supported the alternative view. For instance, in the *De Sapientia Veterum* version of the myth of Cupid, Bacon

[1] Lemmi says of Bacon's interpretation of this myth: 'If I am not mistaken, the symbolism of the essay of Dionysus is, in fact, rather an original development of hints than anything approaching a mosaic of appropriations' (op. cit., p. 202). As I have pointed out before, Bacon is able to sustain parallels in terminology and phrasing between external causal relationships of natural phenomena and of human generation, not because of a conscious mechanistic view of human mental and psychological processes, but on the contrary, because of his extreme animistic view of nature. Apparent good is 'mother' of desire; desire is 'nursed' in the mind; the actions of the mind 'labour and limp'; desire is secretive, 'underground' and without shame or restraint; it grows 'bolder and bolder'; and so on.

[2] And in the isolated myth of Cupid in *De Principiis atque Originibus*.

interprets the birth of Cupid from an egg hatched by Night as symbolising the inaccessibility to man of the summary or highest law of nature. In the *De Principiis atque Originibus* (which according to Spedding is a later work), on the other hand, he interprets the same detail as implying the total accessibility to man of the laws of primary matter by the method of exclusions:

> Now that point concerning the egg of Nox bears a most apt reference to the demonstrations by which this Cupid is brought to light. For things concluded by affirmatives may be considered as the offspring of light; whereas those concluded by negatives and exclusions are extorted and educed as it were out of darkness and night. Now this Cupid is truly an egg hatched by Nox; for all the knowledge of him which is to be had proceeds by exclusions and negatives...But it further intimates, that there is some end and limit to these exclusions; for Nox does not sit for ever. And certainly it is the prerogative of God alone, that when his nature is inquired of by the sense, exclusions shall not end in affirmations. But here the case is different; and the result is, that after due exclusions and negations something is affirmed and determined, and an egg laid, as it were, after a proper course of incubation; and not only that Nox lays her egg, but that from this egg is hatched the person of Cupid: that is to say, not only is some notion of the thing educed and extracted out of ignorance, but a distinct and definite notion.
>
> [V, 463, 465]

Here the same point in the myth is used by Bacon in support of two opposed views on the accessibility of natural knowledge, which he may have held at different periods in his life.

It appears that Bacon uses the philosophical myths as a convincing medium for presenting unfamiliar scientific ideas to a popular audience, just as he advocates in discussing this 'method of discourse' in the *De Augmentis* survey of presentation of knowledge. To this end he reduces the clutter of conflicting detail and alternative explanation, which makes the traditional mythographers' works appear so clumsy. Instead, Bacon's interpretations are vivid, consistent, and lack the anecdotal quality of Comes' interpretations. By retaining Comes' familiar renderings of the myths themselves, Bacon possibly gains the advantage of passive acceptance of these by his readers (familiar with the *Mythologiae*),[1] thus strengthening the impression that a

[1] See note p. 180. T. W. Baldwin gives the *Mythologiae* of Comes as a standard school text on mythography in the period. See Baldwin, *William Shakspere's small Latine and lesse Greeke* (Urbana, 1944), I, 421.

scientific explanation which closely matches the details of these long-established myths must of necessity be true.

It is likely that Bacon did believe that the myths of Pan, Coelum and Cupid concealed earlier attempts at understanding the origins of the world and natural order (this view was widely held in the period). He may consequently have considered that these myths provide particularly fortunate vehicles for communicating scientific truths arrived at by rational means, since the seeds of these truths were already concealed within them in some form.[1] At any rate, Bacon appears to take the myths relating to natural philosophy more seriously than the moral and political myths. His interpretations of the former have a forcefulness and directness which is somehow lacking in the latter.[2]

[1] Bacon allows for myths being *both* repositories for the concealed wisdom of past ages, *and* a vehicle for communicating truths arrived at now. See above, p. 172 and [IV, 317]: 'Now this method...was very much in use in the ancient times...And even now, and at all times, the force of parables is and has been excellent.'

[2] Lemmi, Rossi and Carman all find the philosophical myths a more rewarding proposition than the moral and political myths, as is evidenced by the space devoted to each type. Of thirty-one myths in the *De Sapientia Veterum* only five (Pan, Coelum, Proteus, Cupid, Proserpina) are strictly philosophical; eight are concerned in some way with the possibilities for human knowledge (Actaeon and Pentheus, Orpheus, Daedalus, Ericthonius, Deucalion, Atlanta, Prometheus, Sphinx); while the remaining eighteen are moral and political. Yet Lemmi, for example, devotes nearly twice as much space to the 'symbols of scientific speculation' (comprising the first two groups) as to the 'symbols of worldly wisdom'. My own account reveals a similar imbalance! See Lemmi, op. cit.; Carman, op. cit.; P. Rossi, *Francesco Bacone, dalla magia alla scienza* (Bari, 1958), transl. Rabinovitch (London, 1968), chapter III.

The question of Bacon's own beliefs about the relation between the myths of the ancients and the scientific and ethical truths supposedly revealed by their correct interpretation has been much discussed. Spedding [VI, 607–8] thinks that Bacon seriously believed that the myths concealed philosophical truths, and that the *De Sapientia Veterum* sets out in good faith to uncover these. Lemmi inclines to the view that Bacon followed in the tradition of those who attributed serious intention to the classical writing of the myths, but that he admitted that in the final analysis there is no way of determining whether the myths preceded or succeeded the truths hidden in them (op. cit., pp. 41–4). Rossi puts forward an elaborate theory of Bacon's changing attitude towards myth interpretation, documented from his various discussions of parable as 'method' (op. cit., chapter III, pp. 73–134, especially p. 95). He concludes that 'After wavering in the *Advancement of Learning*, in the preface to the *De Sapientia Veterum* he finally asserted his allegiance to the allegorical tradition' (p. 96).

Both Lemmi and Carman suggest that the interpretations of earlier mythographers directly influenced Bacon's own philosophical views. That is, that he adopted some of the cosmological views of writers like Comes who had previously expounded philosophical interpretations of some of the myths. Spedding and Rossi, on the other hand, seem to suggest that Bacon thought the myths provided indirect confirmation of his scientific views, derived in other ways.

I myself consider that it is virtually impossible on the strength of the available textual evidence to decide whether Bacon believed that the ancients hid their philosophical insights in myths devised for the purpose, or that the myths were literary works whose interpretations were later added as a didactic vehicle for transmission. Bacon concludes

Whilst it is plausible that Bacon seriously believed truths of natural philosophy to be hidden in certain myths, there can be little doubt that in his moral and political interpretations he makes opportunistic use of the myths to communicate precepts in a persuasive form. I have suggested that Bacon adapts the details and phrasings of these myths to enable him to give these precepts a convincing setting, and one in which they are easily grasped. In these myths Bacon makes open use of the possibilities for word play (desire is 'conceived' by apparent good, 'grows strong' with indulgence, and so on). Erasmus was an enthusiastic advocate of this sort of parabolic paralleling to make moral precepts compelling and easy to retain in the memory. His *Parabolae sive Similia* (1514) gives a collection of fables and comparisons to be used by an author to give emphasis, and add conviction to a point in argument. In the *De Copia* he advocates parable as an important device for strengthening a case, as well as for ornamenting a discourse.[1] Bacon's moral and political parables are, I suggest, extended examples of this technique, and are with reason included by Bacon under the heading of 'parabolical poesy', or fictional presentation to a vulgar audience of difficult precepts.

his preface to the *De Sapientia Veterum* as follows: 'Upon the whole I conclude with this: the wisdom of the primitive ages was either great or lucky; great if they knew what they were doing and invented the figure to shadow the meaning; lucky, if without meaning or intending it they fell upon matter which gives occasion to such worthy contemplations. My own pains, if there be any help in them, I shall think well bestowed either way: I shall be throwing light either upon antiquity or upon nature itself' [VI, 698–9]. At the very least, therefore, Bacon believed that there was an extremely fortunate match between myth and what he saw as the truth.

[1] *De duplici copia verborum ac rerum commentarii duo* (Parisiis, 1512).

11
Exempla

In the last chapter I discussed the use which Bacon makes of an existing form of presentation (myth interpretation) as a 'method' for transmitting unfamiliar knowledge to a popular audience. In the present chapter I consider his use of related devices for the same purpose, on a smaller scale. I also distinguish two ways in which Bacon uses devices involving comparison, example and interpretation, but which because of our own intellectual attitudes we tend to confuse. Bacon sometimes uses comparison and analogy of a sort which we regard as extremely far-fetched, as a serious stage in the development of a scientific investigation. In such circumstances, argument by analogy and from example should, I suggest, be considered quite separately from comparison as a vehicle for presenting existing knowledge in readily accessible form, by means of a carefully contrived parallel.

Late-sixteenth-century rhetoric manuals group parable, comparison and *exemplum* together, as using similar means to present an appealing and persuasive argument. All these devices, in different but related ways, both embellish a work, and increase its persuasive force. Here is Puttenham's account of *resemblance* or *similitude* as an instrument of ornament and persuasion:

As well to a good maker and Poet as to an excellent perswader in prose, the figure of *Similitude* is very necessary, by which we not onely bewtifie our tale, but also very much inforce and inlarge it. I say inforce because no one thing more preuaileth with all ordinary iudgements than perswasion by *similitude*. Now because there are sundry sorts of them, which also do worke after diuerse fashions in the hearers conceits, I will set them all foorth by a triple diuision... and I will cal him by the name of *Resemblance* without any addition, from which I deriue three other sorts: and giue euery one his particular name, as *Resemblance* by

Pourtrait or Imagery, which the Greeks call *Icon, Resemblance* morall or misticall, which they call *Parabola,* and *Resemblance* by example, which they call *Paradigma.*[1]

Puttenham goes on to characterise the types of resemblance:

But when we liken a humane person to another in countenaunce, stature, speech or other qualitie, it is not called bare resemblance, but resemblaunce by imagerie or pourtrait, alluding to the painters terme, who yeldeth to th' eye a visible representation of the thing he describes and painteth in his table...

[1] G. Puttenham, *The Arte of English Poesie* (London, 1589), p. 201. Erasmus links fable and comparison as species of 'exemplum' as follows: 'However, most powerful for proof, and therefore for copia, is the force of *exempla,* which the Greeks call παραδείγματα. These are employed either as similes, or *dissimilia,* or contraries, also, in comparing the greater to the lesser, the lesser to the greater, or equals to equals. Dissimilarity and inequality are in class, measure, time, place, and most of the other circumstances that we have enumerated above. This class embraces the *fabula,* the apologue, the proverb, judgments, the parable, or *collatio,* the *imago* and analogy, and other similar ones. And indeed, most of these are customarily used not only for producing belief but also for embellishing and illustrating, for enriching and amplifying subject matter.' *De Copia,* transl. King and Rix (Wisconsin, 1963), p. 67.

Puttenham distinguishes *auricular* (appealing to the ear), *sensable* (appealing to the mind) and *sententious* figures in rhetoric. The last of these divisions contains such literary devices as both please the ear and in an extended fashion further the argument. He discusses parable, comparison and *exemplum* in this last division. Metaphor and allegory fall under *sensable* figures.

On the other hand, Peacham makes a more usual division, following Susenbrotus, thus:

A figure is divided into	Tropes	Words Sentences
	Rhetorical Schemes	Words Sentences Amplifying

where a *trope* plays on the meaning of a word or extended passage, and a *scheme* plays on arrangement either of single words, or of extended passages, or contrived manners of presentation. Comparison, similitude and example all come under the last division, as both elaborating verbally and elucidating intellectually (whilst metaphor and allegory fall under tropes of words and sentences respectively). See H. Peacham, *The Garden of Eloquence* (London, 1593), ed. Crane (Florida, 1954). Peacham's examples show how closely resemblance and example come to one another (examples may be true or feigned, pp. 186–9).

J. Hoskins also places comparison under amplification, and also links example and resemblance. *Directions for Speech and Style* (1599?), ed. Hudson (Princeton, 1935), pp. 17–22. Under 'varying' of discourse Hoskins groups metaphor, similitude, fable, 'poet's tale' (parable in Bacon's sense); pp. 8–10.

Cicero grouped similitude (*imago*), parallel (*collatio*) and example (*exemplum*) under means of supporting an assertion (*De Inventione* I, XXX, 49). This, of course, is in the context of supporting legal cases to produce a judgment in a new instance. Quintilian, discussing types of material to be adduced in support of a question, said: 'Roman writers have for the most part preferred to give the name of comparison to that which the Greeks style παραβολή, while they translate παράδειγμα by example, although the latter involves comparison, whilst the former is of the nature of an example. For my own part, I prefer to regard both as παραδείγματα and to call them examples'. *Institutio Oratoria* V.xi.1 (Loeb translation by H. E. Butler).

In general, in discussion of techniques for supporting a thesis, authors link parable, comparison and example. In discussing *how* to play on resemblance they treat them separately.

And this maner of resemblaunce is not onely performed by likening of liuely creatures one to another, but also of any other naturall thing, bearing a proportion of similitude, as to liken yealow to gold, white to siluer, red to the rose, soft to silke, hard to the stone and such like.[1]

General resemblance or 'bare similitude' and 'Icon' are (judging from Puttenham's examples) primarily ornamental. For more serious purposes *parabola* and *paradigma* are appropriate:

But whensoeuer by your similitude ye will seeme to teach any moralitie or good lesson by speeches misticall and darke, or farre fette, vnder a sence metaphoricall applying one naturall thing to another, or one case to another, inferring by them a like consequence in other cases the Greekes call it *Parabola,* which terme is also by custome accepted of vs: neuerthelesse we may call him in English the resemblance misticall: as when we liken a young childe to a greene twigge which ye may easilie bende euery way ye list: or an old man who laboureth with continuall infirmities, to a drie and dricksie oke...

Finally, if in matter of counsell or perswasion we will seeme to liken one case to another, such as passe ordinarily in mans affaires, and doe compare the past with the present, gathering probabilitie of like successe to come in the things we haue presently in hand: or if ye will draw the iudgements precedent and authorized by antiquitie as veritable, and peraduenture fayned and imagined for some purpose, into similitude or dissimilitude with our present actions and affaires, it is called resemblance by example.[2]

Similitude, icon, parable and example are all techniques for providing affective support by comparison. It should be noted that in this account Puttenham is scarcely interested in whether comparison is made with an imaginary or a real object or event, nor whether the resemblance is one supposed obvious (as when legal or historical precedents are cited) or is contrived for effect (as when blonde hair is compared to gold). To support an argument, all that is necessary is that the comparison should be 'apt', that is that it should be fitting to the occasion, and that there should be a large enough degree of resemblance obvious to the reader for him to accept the resemblance as telling.

Comparison used in any of the ways discussed by Puttenham

[1] Putttenham, op. cit., p. 204. [2] Puttenham, op. cit., p. 205.

does not *demonstrate* the truth of any of the points made. As Puttenham says, it 'inforces' the point; it gives persuasive backing and emphasis.[1] The particular advantage of comparison, as the handbook writers emphasise, is that it is an extremely successful way of providing iterative support in discourse aimed at a 'vulgar' audience: 'no one thing more preuaileth with all ordinary iudgements than perswasion by *similitude*'.[2]

As Bacon makes clear in his incidental comments on comparison, he endorses this traditional view.[3] However, he also advocates another use of comparison as an indication of fruitful lines along which to pursue scientific investigations. Whilst resemblance in this context is still intended to be regarded only as suggestive, it is supposed to contribute to the discovery of knowledge and the interpretation of nature. Resemblances which are used in this way are not merely similitudes (resemblances contrived or stressed for their supporting value in an argument) but indicators of similarities between natural phe-

[1] It should be remembered that in the period 'prove' meant 'provide support for' (or more rarely, test, or put on trial). 'Demonstrate' is the term used for formal proof. See for instance T. Wilson, *Arte of Rhetorique* (1553): 'those that delight to *prove* things by similitudes must learn to know the nature of the things compared' (my italics) (cit. B. W. Vickers, *Francis Bacon and Renaissance Prose* (Cambridge, 1968), p. 147). Whilst Vickers and Tuve are right to emphasise the extremely self-conscious use of comparison in the period, they are mistaken, I feel, in using the term 'logical' to describe such usage. Whilst they aimed at quite specific persuasive effects, writers in the period followed the precepts of formal *rhetoric*, which were believed to make use of the audience's affections, and to work on them in specific ways to procure assent to a particular attitude or approach to some question. They did not claim to be using the techniques of *logic* by which the reason manipulates propositions to arrive at conclusions consistent with the premises of the argument. See R. Tuve, *Elizabethan and Metaphysical Imagery* (Chicago, 1947), passim; Vickers, op. cit., pp. 146–52.

[2] Puttenham, op. cit., p. 201. See also Peacham's commendation of the use of metaphor: 'Apt Metaphors haue their manifold frutes, and the same both profitable and pleasant, which is a thing well known to men of learning and wisedome. First, they giue pleasant light to darke things, thereby remouing unprofitable and odious obscuritie. Secondly, by the aptnesse of their proportion, and nearenesse of affinitie, they worke in the hearer many effects, they obtaine allowance of his iudgement, they moue his affections, and minister a pleasure to his wit. Thirdly, they are forcible to perswade. Fourthly to commend or dispraise. Fiftly, they leaue such a firme impression in the memory, as is not lightly forgotten' (op. cit., p. 13). Of example he says that 'their use in doctrine [teaching] is to be greatly commended, so be it, that they be aptly applyed and truely expressed, for they instruct plainly, moue mightily, and perswade effectually' (p. 188). See also Wilson, op. cit. (1560 ed.), fo. 88r: 'Neither can anye one perswade effectuouslye, and winne men by weyght of his oration, withoute the helpe of woordes altered and translated [transposed]'.

[3] See Vickers, op. cit., pp. 151–3 for a concise summary of Bacon's comments on the persuasive use of comparison. It should however be noted that when Bacon says of the 'schoolmen' that they should have relied on 'evidence of truth *proved by* arguments, authorities, similitudes, examples' [III, 286] (cit. Vickers, op. cit., p. 152; my italics), he means only that similitudes are amongst the illustrative supports for general propositions, and makes the dubious point that such general statement and illustration is a superior technique to quibbling (debating) over artificial detail.

nomena and processes which arise out of the economy and regularity of nature [III, 320; IV, 339]. We have already encountered some of the ways in which Bacon used resemblances between natural processes to suggest profitable directions which investigations could take.[1] Amongst the prerogative instances which help to cut short the stages of the interpretation of nature[2] we find further examples of Bacon's commitment to resemblance as the basis for serious scientific judgments.

Instances conformable, or of *analogy,* which Bacon also calls *parallels* or *physical resemblances,* are prerogative instances which aid investigation by 'exalting the understanding and leading it to genera and common natures' [IV, 247, 164]. Bacon believed that there are certain physical resemblances between natural bodies which are indicative of more far-reaching similarities in their properties. Surface similarity between two bodies may mean that they will also behave similarly under similar circumstances, or that the two bodies were generated, or came into being in similar ways. Thus physical similarity may be an indicator for the fact that the same general rules of behaviour hold for bodies in disparate instances. The similarity in shape (as Bacon believed) between the continents of Africa and South America indicates that the two continents were formed by the same physical process [IV, 167];[3] and the physical similarity between the inner ear and a cave which reflects an echo points to the fact that reflection of sound and reception of sound by the ear are similar processes [IV, 164].[4]

Physical conformity is not an indication that bodies have similar *forms,* since the two bodies whose external features resemble one another may differ radically in the way in which their fundamental parts are put together:

[1] See the discussion of *philosophia prima* above, pp. 101–6 and 159–61.

[2] See above, chapter 6, pp. 124–6.

[3] It has been suggested that Bacon anticipated the theory of continental drift with his discussion of the 'conformity' between Africa and South America. However, as Carozzi has pointed out, Bacon believed Africa and South America to be *alike in shape* not to *fit together* (revealing their origin as one land mass before splitting and drifting). See A. V. Carozzi, 'New historical data on the origin of the theory of continental drift', *Geological Society of America Bulletin,* 81 (1970), pp. 283–6; 'A propos de l'origine de la théorie des dérives continentales: Francis Bacon (1620), Francois Placet (1668), A. von Humboldt (1801) et A. Snider (1858)', *Compte rendu des séances de la société de physique et d'histoire naturelle de Genève,* 4 (1969), pp. 171–9.

[4] Such similarities reveal common processes which are part of the study of *latent process.* See above, p. 142.

[Conformable instances] are those which represent the resemblances and conjugations of things, not in Lesser Forms...but merely in the concrete. Hence they may be called the first and lowest steps toward the union of nature. Nor do they constitute any axiom immediately from the beginning, but simply point out and mark a certain agreement in bodies. But although they are of little use for the discovery of forms, they nevertheless are very serviceable in revealing the fabric of the parts of the universe, and anatomising its members; from which they often lead us along to sublime and noble axioms, especially those which relate to the configuration of the world rather than to simple forms and natures. [IV, 164]

Conformable instances aid the investigations of natural philosophy at a more elementary level than the induction itself. They lead to tentative grouping of bodies as susceptible to similar treatment, in the field of physics.

The investigator must, however, be sure that the resemblances he picks out are genuinely instructive, and not superficial. Bacon rejects the sort of resemblances which were invoked in the doctrine of signatures, and popularised by writers like della Porta.[1] In the *Historia Vitae et Mortis* he casts scorn on such arguments:

But when on the other side I hear discourses on medicines prepared from gold (because forsooth gold is not subject to corruption); on the use of precious stones to refresh the spirits, by reason of their secret properties and brilliancy; that if balsams and the quintessences of living creatures could be received and detained in vessels, there would be good hope of immortality; that the flesh of serpents and deer by a kind of sympathy have power to renew life, because the one casts its slough, the other its horns (they should have added likewise the flesh of the eagle, for the eagle casts its beak);...when I hear of...such like fables and superstitions, I wonder exceedingly that men should be so demented as to be imposed upon by them. [V, 265][2]

[1] G. B. della Porta, *Magia Naturalis* (Neapolis, 1589). For an account of the doctrine of signatures adequate for the present discussion see A. Arber, *Herbals: Their Origin and Evolution* (Cambridge, 1953).

[2] The author Bacon appears to have directly in mind in this denunciation of argument from signatures is Roger Bacon. Spedding points out that the examples above, and others in the same passage, are drawn from Roger Bacon's *De Mirabil. Potest. Artis et Naturae.* This work appeared in an edition of natural philosophy tracts printed in Paris in 1542. Dee produced an English translation in 1618. See [II, 98, 158]. We have here direct evidence of Bacon's familiarity with at least one of Roger Bacon's works (most of his works were still only available in manuscript) – one published in a fashionable collection of tracts of the sort which someone of Bacon's background would be likely to prefer to highly specialist and technical scientific works.

These are accidental resemblances. So also is that which traditionally leads to the attributing of aphrodisiac powers to the herb *satyrion*, because of the resemblance of its root to the male testes. In fact, as Bacon notes, it is an accident of growth that the previous year's tuber remains attached to the plant [IV, 260].[1]

As well as believing that physical resemblances are in limited respects significant, it will be remembered that Bacon considered that the recurrence in different branches of knowledge of universal principles was important for discovering the underlying regularities of nature, and for extending knowledge acquired in one field to others.[2] He identified these universal principles on the basis of similar forms of words: the rule, *A discord ending immediately in a concord sets off the harmony*, applies both in music and in ethics [IV, 339]. That is, a rule employing these terms holds in diverse branches of knowledge, thus indicating the true generality of its application. The most striking of these general rules identified on the strength of repeated forms of words is that concerning putrefaction of bodies: *Putrefaction is more contagious before than after maturity* [IV, 338]. This Bacon holds to apply both in physics and in politics, on the assumption that the use of the term *putrefaction* to describe the deterioration of public morals is comparable with that describing decay of vegetation.[3] Whilst he concedes that such resemblances might seem to be merely verbal, Bacon attributes a deeper significance to them:

> And these are no allusions but direct communities, the same delights of the mind being to be found not only in music, rhetoric, but in moral philosophy, policy, and other knowledges. [III, 230]

Thus Bacon acknowledges the importance of two sorts of natural resemblance in the early stages of all philosophical investigations. Physical similarity in objects may indicate a restricted similarity between further properties, and possibly similar origins and development. Resemblances between processes in different fields of investigation may indicate that both areas can be considered to be governed by the same collection of general

[1] See also [II, 461]: 'They have a foolish tradition in magic, that if a chameleon be burnt upon the top of an house, it will raise a tempest; supposing (according to their vain dreams of sympathies,) because he nourisheth with air, his body should have great virtue to make impression upon the air.'

[2] See the discussion of *philosophia prima*, pp. 101–8 and 159–61.

[3] See above p. 105.

rules, and thus that rules or principles commonly accepted as holding in one field may safely be assumed to hold in the other. The great advantage of such comparative techniques is that they are simple and immediate.[1] As we saw, Bacon rests his entire inductive method on the ability of the investigator to spot the most fundamental natural resemblances.[2] Subsequently the inductive method is supposed to restrain the investigator from basing arguments by analogy on unsound resemblances, and to allow him to make the most of such resemblances as he detects.

The main point to note about such scientific or philosophical use of comparison is that it is extremely tentative. Although, as I have emphasised, Bacon is sometimes persuaded by verbal similarities to consider two processes as comparable, all such resemblances are merely intended to be suggestive of fruitful paths which future investigation might take. When, on the other hand, he uses comparison to reinforce and develop an argument in a literary or didactic work, it is not in this exploratory way. The resemblance is firmly stated, and strengthened 'with words and figures suitable to the occasion' (as Erasmus counselled), so that 'some details are purposely made light of, others are emphasized'.[3]

However, there is an obvious advantage to be gained by using as incontrovertible in a literary context resemblances which have been judged suitable for providing guidance in philosophical investigations. The possibilities for verbal exploration and elaboration are bound to be greater in these cases than where the comparison is more contrived. It is a feature of Bacon's literary or didactic (in Agricola's broad sense) use of comparison which has occasioned comment, that his extended comparisons tend to make use of resemblances which are also appealed to tentatively within natural philosophy.[4] In discussing Bacon's use of comparison or (in Erasmus' broad sense) *exemplum*, in works designed to transmit knowledge persuasively I begin with some of the more frequently used of such 'genuine' comparisons.

[1] Comparison is simple because it involves simple assent to 'images' projected and juxtaposed by the imagination (the repository of images). This was a traditional view.

[2] 'Since there is in natural objects a promiscuous resemblance one to another, insomuch that if you know one you know all...' [V, 505]. See above, chapter 4. Bacon does warn against too hasty recourse to resemblance: 'The human understanding is moved by those things most which strike and enter the mind simultaneously and suddenly, and so fill the imagination; and then it feigns and supposes all other things to be somehow, though it cannot see how, similar to those few things by which it is surrounded' [IV, 56].

[3] *De Copia*, ed. cit., p. 74.

[4] See Vickers, op. cit., pp. 152–4.

Bacon's most consistently exploited image groups are concerned with building processes, journeying processes, processes involving water, processes involving light, and processes involving natural growth.[1] Part of the strength of these sources of comparison lies in their *lack* of originality; they are sources which have deep roots in tradition, for instance, in the Bible and its various translations.[2] The advantage of such comparisons is that they are immediately acceptable to the reader, and allow the author to exploit the less obviously apt or unexceptionable correspondences in detail, by astute word play. As we saw, this is a technique which Bacon uses successfully in his adaptation of the traditional mythographers' interpretations of classical myths. As in the case of the myths, the persuasive force of the comparison depends to some extent on the conviction with which the resemblance is presented, and the confidence with which it is explored. And in the case of extended comparison, as in the case of the myths, I think that we must allow that the basic resemblances are to be regarded as more than accidental.

The link between physical light and intellectual or spiritual 'illumination' is a traditional one. Aquinas had the following to say about the customary association of 'enlightenment' and 'clarity' produced by natural light, and that produced in the mind by spiritual knowledge:

Whether *light* is used in its proper sense in speaking of spiritual things?

In its primary meaning, [light] was intended to signify that which produces clarity in the sense of sight; afterwards it was extended to that which produces clarity in any sort of knowledge. If then, the word *light* is taken in its primary meaning, it is to be understood metaphorically when applied to spiritual things, as Ambrose says. But if it be taken in its common usage, as applied to clarity of every kind, it may properly be applied to spiritual things.[3]

[1] I follow Vickers, op. cit., chapter 6, 'Philosophy and Image-patterns'. W. R. Davis has also considered some image groups in the *Novum Organum*. See Davis, 'The imagery of Bacon's late work', *Modern Language Quarterly*, 27 (1966), pp. 162–73.

[2] See for instance C. S. Lewis's comments on the influence of biblical imagery on the habits of comparison in English literature; *The Literary Impact of the Authorised Version* (London, 1950). Erasmus uses images which could be considered to belong to the same image groups as those Vickers lists for Bacon, but there are, as far as I can ascertain, no detailed parallels. See *Parabolae sive similia* (Argentorati, 1514).

[3] *Summa Theologica*, Q.67, in A. C. Pegis ed., *Basic Writings of Saint Thomas Aquinas* (New York, 1944), I, 630. See also, Philo, *De Opificio Mundi* (which Bacon cites from elsewhere): 'Special distinction is accorded by Moses to life-breath and to light. The one he entitles the "breath" of God, because breath is most life-giving, and of life God is the author, while of light he says that it is beautiful pre-eminently (Gen. i. 4): for the intelligible as far

We may compare this with the following passage in Bacon's *Advancement of Learning:*

To descend from spirits and intellectual forms to sensible and material forms; we read the first form that was created was light, which hath a relation and correspondence in nature and corporal things, to knowledge in spirits and incorporal things. [III, 296][1]

This passage is part of Bacon's theological evidence for the superiority of wisdom to power in the works of creation. Since light corresponds to knowledge, the prior creation of light by God in the order of creation is evidence for the superiority of knowledge in spiritual affairs.

In the essay 'Of Truth' Bacon plays on this association again:

The first creature of God, in the works of the days, was the light of the sense; the last was the light of reason; and his sabbath work ever since, is the illumination of his Spirit. First he breathed light upon the face of matter or chaos; then he breathed light into the face of man; and still he breatheth and inspireth light into the face of his chosen. [VI, 378]

Here he uses it to link physical light both with knowledge and with spiritual (or moral) enlightenment. In the context of the essay he in this way achieves a transition from discussion of 'truth' as adhering to fact, to 'truth' as goodness or spiritual awareness.[2] The passage quoted thus provides a persuasive transition to further the rhetorical development of the essay.

In the *De Augmentis* Bacon extends the association of physical light and knowledge to the processes of transmission:

The object of philosophy is threefold – God, Nature, and Man; as there are likewise three kinds of ray – direct, refracted, and reflected. For nature strikes the understanding with a ray direct; God, by reason of the unequal medium (viz. his creatures), with a ray refracted; man, as shown and exhibited to himself, with a ray reflected. Philosophy may

surpasses the visible in the brilliancy of its radiance, as sunlight assuredly surpasses darkness and day night, and mind, the ruler of the entire soul, the bodily eyes. Now that invisible light perceptible only by the mind has come into being as an image of the Divine Word who brought it within our ken: it is a super-celestial constellation, fount of the constellations obvious to the sense' (Loeb edition, transl. Colson and Whitaker, 30).

1 In the *De Augmentis* this passage reads: 'A Spiritibus et Intelligentiis ad formas sensibiles et materiatas descendentes, legimus primam formarum creatarum fuisse Lucem; quae in naturalibus et corporeis, Scientiae in spiritualibus atque incorporeis respondet' [I, 465]. Bacon is referring, of course, to Genesis i, 1: 'In the beginning God created the heaven and the earth. And the earth was without form, and void; and darkness was upon the face of the deep. And the Spirit of God moved upon the face of the waters. And God said, let there be light: and there was light.'

2 For detailed discussion of this essay see below, chapter 13, pp. 244–8.

therefore be conveniently divided into three branches of knowledge: knowledge of God, knowledge of Nature, and knowledge of Man, or Humanity. [IV, 337]

In this case the comparison of light with knowledge serves two purposes. It makes vivid the conventional partition of philosophy,[1] and at the same time it reinforces the idea on which Bacon's *philosophia prima* rests – that the *substance* of all branches of philosophy is the same, and only the manner of operation of the basic laws varies with the branch of philosophy. This same use is made of the comparison in the opening book of the *De Augmentis* (and Book 1 of the *Advancement of Learning*):

The essential form of knowledge...is nothing but a representation of truth: for the truth of being and the truth of knowing are one, differing no more than the direct beam and the beam reflected. [III, 287]

In all these cases it is reasonably clear that the 'light' analogy is of persuasive force in sustaining an intellectual point. Bacon's repeated use of the comparison suggests that he regarded it as particularly apposite.[2] But various remarks in ostensibly serious scientific passages of the *De Augmentis* point to the fact that Bacon may have attributed some of the rhetorical success of the comparison to a stronger link between light and knowledge. In

[1] See the discussion of scholastic classifications of knowledge in chapter 4. The partition does not depend upon Bacon's comparison, but is supported by it. In the *Advancement of Learning* the partition is made without explicit use of the 'light' analogy: 'In Philosophy, the contemplations of man do either penetrate unto God, or are circumferred to Nature, or are reflected or reverted upon Himself. Out of which several inquiries there do arise three knowledges, Divine philosophy, Natural philosophy, and Human philosophy or Humanity' [III, 346]. The analogy is introduced later, at the end of the discussion of natural philosophy: 'Thus have we now dealt with two of the three beams of man's knowledge; that is *Radius Directus*, which is referred to nature, *Radius Refractus*, which is referred to God...There resteth *Radius Reflexus* whereby Man beholdeth and contemplateth himself' [III, 366]. Note the reversal of the terms in the comparison.

[2] In the *Sylva Sylvarum* this analogy is used to make the following point: 'The eye of the understanding is like the eye of the sense; for as you may see great objects through small crannies or levels, so you may see great axioms of nature through small and contemptible instances' [II, 377]. And in the *New Atlantis* the analogy is used as follows: 'That every twelve years there should be set forth out of this kingdom two ships, appointed to several voyages; That in either of these ships there should be a mission of three of the Fellows or Brethren of Salomon's House; whose errand was only to give us knowledge of the affairs and state of those countries to which they were designed, and especially of the sciences, arts, manufactures, and inventions of all the world; and withal to bring unto us books, instruments, and patterns in every kind...thus you see we maintain a trade, not for gold, silver, or jewels, nor for silks; nor for spices; nor any other commodity of matter; but only for God's first creature, which was *Light*: to have *light* (I say) of the growth of all parts of the world' [III, 146–7]. See also 'experiments of *light*': '[the old science] has sought, I say, experiments of Fruit, not experiments of Light; not imitating the divine procedure, which in its first day's work created light only and assigned to it one entire day' [IV, 17].

Book 4 of the *De Augmentis* his discussion of the 'form of light' as an essential and overlooked component in the consideration of 'Sense and the Sensible' [IV, 401] suggests that he regarded light as a vital ingredient in all considered action. He appears to believe that all physical interaction between bodies constitutes 'perception', but that only with the reception of light from the object is that object 'sensed' [IV, 403–4]. If Bacon did believe that light carries some visible emanation from the source which is responsible for our conscious experience of an object, then his light analogy is confidently used for its true appropriateness.[1] And this goes also for the extensive play which Bacon makes on the analogy between the mind receiving knowledge and a mirror reflecting light from its surface.[2] The analogy between a mirror and the eye figures amongst the serious resemblances in process in *philosophia prima* [IV, 399].[3]

Bacon uses more elaborate rhetorical play on similarity of process in the group of images connected with water. In the *De Augmentis* the division of knowledge into that acquired from divine and from natural sources is supported by a comparison with sources of water as follows:

> The knowledge of man is as the waters. Some waters descend from above, and some spring from beneath; and in like manner the primary division of sciences is to be drawn from their sources; of which some are above in the heavens, and some here below. For all knowledge admits of two kinds of information; the one inspired by divine revelation, the other arising from the senses. For as to that knowledge which man receives by teaching, it is cumulative and not original; as it is likewise in waters, which beside their own springheads, are fed with other springs and streams. [IV, 336]

The implication of this comparison is that the matter of the sciences is uniform, like water, and differs only in the manner in which it joins the 'well' of all knowledge. Moreover, it emphasises the secondary and subsidiary nature of teaching (what is already known or 'stored') – knowledge acquired from a teacher is 'cumulative and not original'. In an earlier passage this uniformity of knowledge, and 'collecting and storing' character of teaching, are further played upon:

[1] Such a belief was a traditional Aristotelian one.

[2] Vickers collects these images together, op. cit., pp. 189–91.

[3] See also [II, 434; IV, 165].

For as water, whether it be the dew of Heaven or the springs of the earth, easily scatters and loses itself in the ground, except it be collected into some receptacle where it may by union and consort comfort and sustain itself (and for that cause the industry of man has devised aqueducts, cisterns, and pools, and likewise beautified them with various ornaments, for magnificence and state as well as for use and necessity); so this excellent liquor of knowledge, whether it descend from divine inspiration or spring from human sense, would soon perish and vanish into oblivion, if it were not preserved in books, traditions, and conferences; and especially in places appointed for such matters, as universities, colleges, and schools, where it may have both a fixed habitation and means and opportunity of increasing and collecting itself. [IV, 284–5]

In the first passage the comparison is used primarily to establish the partition of knowledge; in the second it supports an argument by the verbal parallels drawn between the two areas. In both cases the source of comparison is introduced first, selecting for emphasis those details which will be necessary to sustain the point of the comparison. The intellectual point is then made, couched as closely as possible in terms of (in this case) the collecting and storing of natural water supplies. This procedure is, in fact, very close to that which I discussed above in connection with Bacon's use of parable: the close match between the interpretation and the deliberately selective description conveys a sense of precision and incontrovertibility. As in the case of the parables, comparisons of this sort possess the persuasive advantage that 'ideas that are objects of the intellect are represented in forms that are objects of the sense' [IV, 315].

In the case of the image of dew and natural springs for divine and natural knowledge, the match is seen to be particularly close if we take into account some of Bacon's physical beliefs:

It seemeth that there be these ways (in likelihood) of version [transmutation] of vapours or air into water and moisture. The first is cold; which doth manifestly condense; as we see in the contracting of the air in the weather-glass; whereby it is a degree nearer to water. We see it also in the generation of springs, which the ancients thought (very probably) to be made by the version of air into water, holpen by the rest which the air hath in those parts; whereby it cannot dissipate; and by the coldness of rocks; for there springs are chiefly generated. We see it also in the effects of the cold of the middle region (as they call it) of the air; which produceth dews and rains. [II, 348]

Dew is air from the heavens transmuted by cold; springs are air from the earth congealed. Dews and springs are therefore associated with rarified, spirituous matter, as is knowledge (with the substance of the animal spirits). It seems possible here, as in the case of the light/knowledge comparison, that Bacon regards the analogy as a telling natural one.

It would not do to overstate the scientific foundation for Bacon's literary imagery. It is, however, fair to say that amongst Bacon's 'vocabulary' of images, used repeatedly to support argumentative points, some at least are linked with resemblances which he expected to provide insight in natural philosophy.[1] The important difference between Bacon's use of comparison in his scientific investigations and his use of comparison (even the same ones) for transmission of knowledge, is that scientific resemblances are used tentatively, as a guide to more solid investigation, whilst literary comparison is exploited for all it is worth.

Apophthegm

At the beginning of this chapter I emphasised the close relation, in the period in which Bacon wrote, between parable, comparison, apophthegm (or witty saying), proverb and example. In the remainder of this chapter I consider the last three of these devices as used by Bacon, so as to complete this discussion of his literary use of *exemplum* (in Erasmus' broad sense) as a 'method' for transmitting unfamiliar knowledge to an untrained audience.

Apophthegms are brief, witty comparisons, often relying on play on words. They do not seem to be intended to point to serious resemblances, significant for scientific investiga-

[1] Various authors have grouped recurring images in Bacon's work. B. V. M. Bowman, *The English Prose Style of Sir Francis Bacon* (unpublished Ph.D. diss., Wisconsin, 1964) picks out six groups: 'clowdes of error', 'natur's warehouse', 'truth barren', 'web of wit', 'happie match', 'thrall to nature' ('Patterns of Imagery', pp. 109–40). Davis, art. cit., discusses light images, circle images, architecture images, mirror images, model images, priest and marriage images. Vickers, op. cit., p. 175, picks out images 'of building and voyaging, images of natural growth, images involving water, and images of light'. He discusses theatre images in 'Bacon's use of theatrical imagery', *Studies in the Literary Imagination*, 4 (1971), pp. 189–226.

From these studies it is clear that Bacon uses a compact range of images for repeated effect (Bowman is probably right in thinking that Bacon kept a commonplace book of images appropriate to particular themes, op. cit., p. 82).

tions.[1] Here are two of Bacon's collection of apophthegms, attributed to Mr Bettenham, Reader at Gray's Inn:

Mr. Bettenham used to say; *That riches were like muck; when it lay upon an heap, it gave but a stench and ill odour; but when it was spread upon the ground, then it was cause of much fruit.*
The same Mr. Bettenham said; *That virtuous men were like some herbs and spices, that give not their sweet smell, till they be broken and crushed.*
[VII, 160]

Witty comparisons of this sort are used by Bacon to give strong emphasis to a point put forward in the course of discussion.[2] The first of the apophthegms quoted above is used in the essay 'Of Sedition and Troubles' to support the precept that it is inadvisable to allow the wealth of a nation to be controlled by a small number of officials:

Above all things, good policy is to be used that the treasure and monies in a state be not gathered into few hands. For otherwise a state may have a great stock, and yet starve. And money is like muck, not good except it be spread. This is done chiefly by suppressing, or at the least keeping a straight hand upon the devouring trades of usury, ingrossing, great pasturages, and the like. [VI, 410]

The second occurs in the essay 'Of Adversity' to support the theme that adversity should be regarded as a welcome test of fortitude and divine mercy:

Certainly virtue is like precious odours, more fragrant when they are incensed or crushed: for Prosperity doth best discover vice, but Adversity doth best discover virtue. [VI, 386]

Bacon, like many of his contemporaries, collected in a notebook apophthegms which struck him in his reading, or which

[1] Bacon does of course use many illustrative comparisons which whilst not based on 'real' resemblances are not so contrived as to qualify as purely witty apophthegms or puns. The following is an example of such intermediate use of comparison: 'certain it is that words, as a Tartar's bow, do shoot back upon the understanding of the wisest, and mightily entangle and pervert the judgment' [III, 396]. I have not discussed such comparisons. They appear to be of the strictly traditional type such as Erasmus collects together in his *Parabolae sive similia*.

[2] Apophthegms were traditionally used for illustration in this way. See for example Erasmus, *Adagia* (Venetiis, 1508). On use of apophthegms see R. R. Bolgar, *The Classical Heritage and its Beneficiaries* (Cambridge, 1958), p. 298: 'To be acceptable [as a rhetorical illustration], an adage [Bacon's 'apophthegm'] must be well known: a popular saying, a quotation from poetry or drama, an historical reference or a technical phrase from a common craft or profession. Furthermore, it must be striking, either in its subject-matter or in its structure, a quality generally attained by metaphor.'

were reported to him, or heard by him in conversation with others.[1] These were to provide illustration and support for points in discourse:

Neither are Apophthegms themselves only for pleasure and ornament, but also for use and action. For they are (as was said) 'words which are goads,' words with an edge or point, that cut and penetrate the knots of business and affairs. Now occasions are continually returning, and what served once will serve again; whether produced as a man's own or cited as an old saying. [IV, 314]

In its original context the apophthegm is a spontaneous, witty remark, which goes to the heart of the matter in hand, and by its form does not give offence.[2] When borrowed and exploited by another author (sometimes with a brief account of the original accompanying incident as well), the apophthegm is a supporting device which is intended to drive home a new point:

Certainly they [apophthegms] are of excellent use. They are *mucrones verborum, pointed speeches.* Cicero prettily calls them *salinas, saltpits*; that you may extract salt out of, and sprinkle it where you will. They serve to be interlaced in continued speech. They serve to be recited upon occasion of themselves. They serve if you take out the kernel of them, and make them your own. [VII, 123]

Thus like parable and comparison, the apophthegm has an active and confirmatory rôle to play in discourse. It may positively further the course of the argument, or it may be recalled to confirm and persuade.

Bacon has an original explanation to offer for the peculiar power of the apophthegm to express even insults in a form which amuses, rather than offends, the party at whom it is directed. He maintains that such expressions are exactly comparable with musical phrases which move from discord to harmony, combining painful and pleasurable stimuli of the senses in proportions which result in extreme pleasure:

[1] See *Apophthegms New and Old* [VII, 123–65]. Students of the period were actively encouraged by their teachers to keep by them a notebook in which to record any phrase or passage which they found particularly striking in reading or in conversation. See Erasmus, *De duplici copia verborum ac rerum commentarii duo* (Parisiis, 1512); C. Hoole, *A new discovery of the old arte of teaching schoole* (London, 1660). See also Bolgar, op. cit., pp. 273–5.

[2] See the essay 'Of Sedition and Troubles' [VI, 412]: 'Surely Princes had need, in tender matters and ticklish times, to beware what they say; especially in these short speeches, which fly abroad like darts, and are thought to be shot out of their secret intentions.'

So again a man should be thought to dally, if he did note how the figures of rhetoric and music are many of them the same...Plutarch hath almost made a book of the Lacedaemonian kind of jesting, which joined ever pleasure with distaste. *Sir,* (saith a man of art to Philip King of Macedon when he controlled him in his faculty,) *God forbid your fortune should be such as to know these things better than I.* In taxing his ignorance in his art he represented to him the perpetual greatness of his fortune, leaving him no vacant time for so mean a skill. Now in music it is one of the ordinariest flowers to fall from a discord or hard tune upon a sweet accord. [III, 230]

The observation that Philip knows nothing of the jester's art is potentially insulting, but in the apophthegm it is made pleasant by the conceit that Philip's ignorance is the product of his good fortune. In the *Sylva Sylvarum* Bacon suggests an explanation for the comparable effects of combined discord and harmony in music, and the displeasing and pleasing combination of sentiments in the apophthegm:

Again, the falling from a discord to a concord, which maketh great sweetness in music, hath an agreement with the affections, which are reintegrated to the better after some dislikes; it agreeth also with the taste, which is soon glutted with that which is sweet alone. [II, 389]

...the senses love not to be over-pleased, but to have a commixture of somewhat that is in itself ingrate. Certainly we see how discords in music, falling upon concords, make the sweetest strains. [II, 612]

The force of the apophthegm's effect is not the product of intellectual content, but of the peculiar emotional response it induces. It combines pleasing aptness with possible offence, either of sense (because of its punning play on words) or of the sensibility of the recipient (where it masks a disrespectful sentiment, or unwelcome insight). This explanation places apophthegm in the same bracket as parable and comparison, which prompt intellectual assent by the 'non-rational' activity of assessing juxtaposed images in the imagination.

Proverb

Bacon regards the aphorism as the ideal form of unmisleading transmission of general precepts, amongst 'magistral' methods (which do not rehearse the procedure of discovery of the knowledge communicated).[1] The crucial characteristics of such

[1] See above, chapter 9.

aphorisms are their terseness, their generality, and their open-endedness, or context-free formulation. Amongst the works of the ancients whose presentation fulfils these criteria Bacon includes Solomon's *Proverbs*:

For we see all the ancient wisdom and science was wont to be delivered in that form [aphorisms]; as may be seen by the parables of Solomon, and by the aphorisms of Hippocrates, and the moral verses of Theognis and Phocylides: but chiefly the precedent of the civil law, which hath taken the same course with their rules, did confirm me in my opinion. [VII, 321]

Neither was this of use only with the Hebrews, but it is generally found in the wisdom of the ancients, that as men found out any observation which they thought good for life, they would gather it and express it in some short proverb, parable, or fable. [V, 56]

In his discussion of Solomon's proverbs Bacon makes it clear that one of the incidental advantages of their compressed general formulation is that it lends itself to quotation, and hence to long-term preservation. Succeeding generations can use the proverb or aphorism as a cue to reformulation of its general significance. Bacon himself attempts to do this for some of Solomon's proverbs. For instance:

PROVERB

A poor man that oppresses the poor, is like a sweeping rain, which causes famine.

Explanation

This proverb was anciently figured and represented under the fable of the full and hungry horseleech; for the oppression of a poor and hungry man is far more severe than that of a rich and full one, inasmuch as the former practises all the arts of exactions, and searches every corner for money. The same used also to be likened to a sponge, which when dry sucks in strongly, but not so when wet. And it contains a useful warning for princes and peoples; for princes, that they commit not offices or the government of provinces to needy persons and such as are in debt; for peoples, that they allow not their rulers to be too much in want of money. [V, 50]

Bacon's approach to proverb interpretations appears to be comparable with his approach to myths and their interpretations (see preceding chapter). He believes that the proverbs were originally meant to contain in pithy and striking form the author's wisdom. And he also sees the interpretation of the proverbs or aphorisms of the past as an opportunity for transmitting further insights under cover of the close match between the proverb, the interpretation, and the supporting illustrations. Bacon's interpretations of Solomon's proverbs include political and ethical precepts which involve highly imaginative use of the original text, and go beyond conventional attempts at such exegesis.[1]

But because the proverb is so quotable, and so general in application, it also lends itself to another use. It joins the list of devices which may be called upon to support a point in persuasive discourse. In this case the context for the interpretation of the proverb is provided by the discussion in hand. By restricting the possible associations, the proverb appears to give strong backing for a particular thesis put forward. For example, in the essay 'Of Ceremonies and Respects' a proverb is used for illustrative support as follows:

It is loss also in business to be too full of respects, or to be curious in observing times and opportunities. Solomon saith, *He that considereth the wind shall not sow, and he that looketh to the clouds shall not reap.*
[VI, 501]

Exemplum

In the case of *exemplum* we have Bacon's own account of the double rôle which this device plays, depending on whether it is seriously considered as the basis for generalisation, or used for illustration:

And it contributes much more to practice, when the discourse or discussion attends on the example, than when the example attends upon the discourse. And this is not only a point of order, but of substance also. For when the example is laid down as the ground of the discourse, it is set down with all the attendant circumstances, which may sometimes correct the discourse thereupon made, and sometimes supply it, as a very pattern for imitation and practice; whereas examples alleged for the sake of the discourse, are cited succinctly and without particularity, and like slaves only wait upon the demands of the discourse. [V, 56]

[1] For a more conventional example of proverb interpretation see Melanchthon, παροιμίαι, *sive Proverbia Solomonis...cum adnotationibus Ph. Melanchthonis* (Haganoae, 1525). I have not been able to find any detailed parallels between Melanchthon's treatment and Bacon's, except possibly the common inclusion of the 'casting pearls before swine' illustration in the interpretation of Proverbs ix. 7 ([V, 39]; Melanchthon, op. cit., II, fo. 34r); and the interpretation of 'winds' as 'idle rumours' in Proverbs xi. 29 ([V, 40–1]; Melanchthon, op. cit., II, fo. 47v). On moral interpretation of biblical texts see B. Smalley, *The Study of the Bible in the Middle Ages* (Oxford, 1952).

In chapter 8 I discussed civil history as a source of constructive examples (particular observations derived from historical narrative) which enable the critical reader to formulate generalisations in fields like politics and law, and to adjust those general precepts to suit new cases. Bacon views this use of *exemplum* as of particular practical importance:[1]

for knowledge drawn freshly and in our view out of particulars knows best the way back to particulars again. [V, 56]

In such circumstances the 'discussion attends on the example', and the examples themselves are 'a very pattern for imitation and practice'. As illustrative devices, on the other hand, examples do indeed 'wait upon the demands of the discourse'. They are carefully selected from the author's fund of striking historical instances, to strengthen a position already taken, and to add to the persuasive weight of the argument. In this case the similarity between the case of the example and the case in point is deliberately strengthened by similar forms of words, and by emphasis on the authority of the source from which the example is drawn.[2]

Two examples of historical illustration from the *Essays* will show particularly clearly the difference between Bacon's use of historical *exemplum* as the basis for political and ethical generalisation, and its use to sustain a didactic point. The following example occurs in the 1625 essay 'Of Suspicion':

[Suspicions] are defects, not in the heart, but in the brain; for they take place in the stoutest natures; as in the example of Henry the Seventh of

[1] Several critics have commented on Bacon's use of *exempla* as 'a very pattern for imitation and practice' [V, 56]. G. H. Nadel, 'History as psychology in Francis Bacon's theory of history', *History and Theory*, 5 (1966), pp. 275–87, reprinted in Vickers ed., *Essential Articles for the study of Francis Bacon* (Connecticut, 1968), pp. 233–50, suggests that Bacon 'thought it possible to have precepts properly induced from, and controlled by, examples' (p. 249). D. S. T. Clark suggests that 'Bacon was able to view the historical example as the equivalent of those types of "instances", particularly "Prerogative Instances", which contributed such an important empirical foundation for true induction in Book II of the *Novum Organum*'; *Francis Bacon: The Study of History and the Science of Man* (unpublished Ph.D. diss., Cambridge, 1970), p. 270. Whilst both authors are right to stress that Bacon in this case uses historical *exempla* as more than illustrative support for prior precepts, I do not think it is possible to sustain the claim that *exempla* become 'simple empirical data' for the inductive method.

[2] On the traditional use of *exempla* for support and illustration see J. T. Welter, *L'exemplum dans la littérature religieuse et didactique du moyen âge* (Paris, 1927); R. Pineas, 'The polemical *exemplum* in sixteenth century religious controversy', *Bibliothèque d'Humanisme et Renaissance*, 28 (1966), pp. 393–6; Pineas, 'John Frith's polemical use of rhetoric and logic', *Studies in English Literature*, 4 (1964), pp. 85–100; J. H. Mosher, *The exemplum in early religious and didactic literature in England* (New York, 1911).

England. There was not a more suspicious man, nor a more stout. And in such a composition they do small hurt. For commonly they are not admitted, but with examination, whether they be likely or no? But in fearful natures they gain ground too fast. [VI, 454]

Henry VII is used as an example again in the 1625 expansion of the essay 'Of Honour and Reputation':

In the third place [amongst the degrees of sovereign honour] are *liberatores,* or *salvatores,* such as compound the long miseries of civil wars, or deliver their countries from servitude of strangers or tyrants; as Augustus Caesar, Vespasianus, Aurelianus, Theodoricus, King Henry the Seventh of England, King Henry the Fourth of France. [VI, 506]

Whereas in the *History of Henry VII* Bacon describes Henry's character, and the events of his reign in meticulous (if sometimes speculative) detail, in order to reveal the underlying patterns and pressures at work, here Henry's qualities are stated dogmatically, 'without particularity' [V, 56] and no attempt is made to back up the plain assertions. The examples are intended to give the persuasive support of their bulk and authority. In general examples used in this way are chosen from those standard in classical and contemporary literature (and therefore above question): Henry VII was suspicious and covetous; 'Socrates, Aristotle, Galen, were men full of ostentation' [VI, 504]; 'the battle of Actium decided the empire of the world' [VI, 451]. They provide confirmation for any point to which they are appropriate, simply by virtue of their common currency. In fact, of course, this piling up of examples is an informal rhetorical induction,[1] which secures assent for a point without ever exposing it as problematic, as in the following example:

[1] In the *Topics* Aristotle advocates rhetorical induction as the best way of convincing a vulgar audience (rather than by syllogistic reasoning): 'we must distinguish how many species there are of dialectical [disputational] arguments. There is on the one hand Induction, on the other Reasoning...Induction is a passage from individuals to universals, e.g. the argument that supposing the skilled pilot is the most effective, and likewise the skilled charioteer, then in general the skilled man is best at his particular task. Induction is more convincing and clear: it is more readily learnt by the use of the senses, is applicable generally to the mass of men, though Reasoning is more forcible and effective against contradictious people' (*Topics* I, 12 (105a10-18)); see also (156a4-7; 157a19-20; 164a12). Aristotle also advises that examples should be chosen from well-known sources (157a13). Cicero advocates induction as a polemical tool because it 'wins his approval of a doubtful proposition because this resembles the facts to which he has assented', and advises that the individual examples in the induction should be such that their truth is taken for granted. *De Inventione* I, xxxi.51-xxxiv.57.

When one of the factions is extinguished, the remaining subdivideth; as the faction between Lucullus and the rest of the nobles of the senate (which they called *Optimates*) held out awhile against the faction of Pompey and Caesar; but when the senate's authority was pulled down, Caesar and Pompey soon after brake. The faction or party of Antonius and Octavianus Caesar against Brutus and Cassius, held out likewise for a time; but when Brutus and Cassius were overthrown, then soon after Antonius and Octavianus brake and subdivided. These examples are of wars, but the same holdeth in private factions. [VI, 499]

In Bacon's writings the related literary devices of parable, comparison, apophthegm, proverb and example function in two ways. The parables, apophthegms and proverbs of the ancients may conceal hidden wisdom; in any case they provide a persuasive vehicle for Bacon's own views in natural philosophy, politics and ethics. Comparison and example may provide a basis for tentative generalisations about the regularities of nature and human politics and social dealings; adduced in support of a point already formulated by an author in discourse these devices reinforce and support the argument. In all these cases *resemblance* in its broadest sense provides a way of forming general propositions, and applying pressure in an argument, without recourse to syllogistic. Hence its appeal to audiences untrained in formal reasoning.

12

Bacon's view of rhetoric

As I indicated in chapter 1, the reorganisation of the dialectic handbook in the sixteenth century left only ornamentation and delivery as the peculiar province of rhetoric.[1] I have pointed out in the last three chapters that Bacon considered 'Method of discourse' or the 'Wisdom of Transmission' to play an integral part in the study of presentation of existing knowledge. And I have discussed some of the ways in which he adapted existing 'methods' of presentation, some of which were previously treated by dialecticians, as part of the wisdom of transmission. Nevertheless, Bacon, like other educational reformers, chooses to reserve the term *rhetoric* for 'the doctrine concerning the Illustration of Discourse' [IV, 454], or ornamentation. The end of rhetoric is 'to fill the imagination with observations and images', so that by embellishment and illustration the author's views are made appealing to his audience, without the use of formal argument [IV, 456].

Some of Bacon's incidental comments on music, and on the similarities between music and rhetoric, suggest that he believed the persuasive effectiveness of figures of rhetoric which rely on arrangements of words to be the direct result of the way in which the senses are stimulated:

The repetitions and traductions[2] in speech and the reports and haunt-

[1] See above, p. 33.

[2] 'Then haue ye a figure which the Latines call *Traductio*, and I the tranlacer: which is when ye turne and tranlace a word into many sundry shapes as the Tailor doth his garment, and after that sort do play with him in your dittie: as thus,

> Who liues in loue his life is full of feares,
> To lose his loue, liuelode or libertie
> But liuely sprites that young and recklesse be,
> Thinke that there is no liuing like to theirs.'

(Puttenham, *Arte of English Poesie*, 1589 ed., p. 170). See also Bacon: 'the reports and fuges have an agreement with the figures in rhetoric of repetition and traduction' [II, 389].

ings of sounds in music are the very same things. Plutarch hath almost made a book of the Lacedaemonian kind of jesting, which joined ever pleasure with distaste...Now in music it is one of the ordinariest flowers to fall from a discord or hard tune upon a sweet accord. The figure that Cicero and the rest commend as one of the best points of elegancy, which is the fine checking of expectation, is no less well known to the musicians when they have a special grace in flying the close or cadence [in an interrupted cadence]. [III, 230][1]

Bacon believed that music has a particularly direct effect on the emotions because of its direct stimulation *via* the ear of the spirits which are responsible for states of the affections:

So it is sound alone that doth immediately and incorporeally affect most. This is most manifest in music, and concords and discords in music; for all sounds, whether they be sharp or flat, if they be sweet, have a roundness and equality; and if they be harsh, are unequal; for a discord itself is but a harshness of divers sounds meeting. [II, 561][2]

In traditional accounts the figure of 'checking of speech' is supposed to imitate the natural breaking off of speech through shame, fear or anger, which prevents the speaker from continuing:

Aposiopesis: by which the orator through some affection, as either feare, anger, sorrow, bashfulness or such like, breaketh off his speech before it

[1] See also [II, 388–9]. Puttenham suggests the general similarity between figures which affect the ear and musical cadence, by his use of musical terms to describe rhetorical figures. For example: 'A word as he lieth in course of language is many wayes figured and thereby not a little altered in sound, which consequently alters the tune and harmonie of a meeter as to the eare' (ed. cit., p. 134). It was customary to compare metre and rhyme in poetry with music; Puttenham does not approach Bacon's detailed matching of rhetorical figures with musical figures.

[2] For the direct relation between movements of spirits and states of the affections see [II, 567–71]. For the relation between states of mind and music see [II, 389–90]. D. P. Walker, *Spiritual and Demonic Magic from Ficino to Campanella* (London, 1958), gives an account of Ficino's theory of the effects of music on the animal spirits (pp. 3–11), which shows that other writers shared Bacon's views. From Walker's account it appears that Campanella's beliefs about the effects of music match Bacon's quite closely: 'The marvellous effects of music are due to this transmission of movement from the air to the spirit. But, unlike Ficino and most later musical humanists, Campanella does not think of these effects as primarily ethical or emotional, but as therapeutic or producing sheer pleasure or pain. The human spirit has a natural rhythmic movement, indicated by the pulse, which is essential to its preservation; music which produces in the air movements similar to, but a little stronger than this, will confirm and encourage this natural movement of the spirit. Low sounds bruise, condense and thicken the spirit; high ones rarify and lacerate it; what is required then, is a combination of the two which is "consonant" to the spirit's natural movement. The spirit, being thereby preserved and strengthened, is delighted; hence the pleasure caused by "consonant" music. For Campanella as for Bacon, musical consonance is not determined by the simple mathematical ratios of two or more sound waves or vibrating strings, but is an entirely relative quality determined by the conformity of musical sounds to any given kind of spirit' (p. 231).

be all ended. *Virgil*: 'How doth the childe Ascanius, whom tymely *Troy* to thee –' breaking off by the interruption of sorrow.[1]

But on Bacon's account its effect on the audience is not produced by sympathy for the speaker's state of mind.[2] It is a purely auditory stimulus which produces a particular pleasant feeling in the listener, moving him to agree with the utterance. This explanation leaves room in rhetorical composition for conscious exploitation of a mechanistic theory of figures.[3]

It appears then that when Bacon calls rhetoric 'Imaginative or Insinuative Reason' [III, 383] he is referring to the peculiar mode of action of rhetorical devices. According to his theories, these act either by stimulating images in the imagination which are judged similar enough to sustain a point (as in the case of supporting devices which make use of resemblance or *exemplum*), or by direct sensory stimulation (in the case of rhetorical figures which depend on patterns of words). Both can be used to subvert reason, which ultimately is responsible for the reader's acceptance of an argument as persuasive. These devices sidestep formal inference, and work directly on the subsidiary faculties (the senses and imagination). Bacon defines rhetoric as follows:

Rhetoric is subservient to the imagination, as Logic is to the understanding; and the duty and office of Rhetoric, if it be deeply looked into, is no other than to apply and recommend the dictates of reason to imagination, in order to excite the appetite and will. [IV, 455]

[1] Peacham, *Garden of Eloquence* (London, 1577, 1593), cit. Vickers, *Classical Rhetoric in English Poetry* (London, 1970), p. 109.

[2] Wilson suggests such an effect: 'There is no substaunce of it self, that wil take fire, excepte ye put fire to it. Likewise, no mannes nature is so apt, streight to be heated, except the Orator him self, be on fire, and bring his heate with him...Again nothyng moisteth soner then water. Therefore a wepyng iye causeth muche moysture, and prouoketh teares. Neither is it any meruaile: for suche men bothe in their countenaunce, tongue, iyes, gesture, and in al their body ells, declare an outwarde grief, and with wordes so vehemently and unfeinedly, settes [sic] it forward, that they will force a man to be sory with them, and take part with their teares, euen against his will' (*Arte of Rhetorique* (1553), 1560 ed., fo. 68v).

[3] Puttenham includes *aposiopesis* amongst 'auricular figures' which he describes as follows: 'As your single words may be many waies transfigured to make the meetre or verse more tunable and melodious, so also may your whole and entire clauses be in such sort contriued by the order of their construction as the eare may receiue a certaine recreation, although the mind for any noueltie of sence be little or nothing affected. And therefore al your figures of *grammaticall* construction, I accompt them but merely *auricular* in that they reach no furder then the eare. To which there will appeare some sweete or vnsauery point to offer you dolour or delight, either by some euident defect, or surplusage, or disorder, or immutation in the same speaches notably altering either the congruitie *grammaticall*, or the sence, or both' (ed. cit., pp. 135–6). See also p. 196. This is the nearest I have found to Bacon's view of the 'auricular' effect of figures. Both Puttenham and Bacon emphasise the mechanical effect of figures at the expense of the traditional claim for the imitative and natural emotive force of rhetoric. For an account of the traditional emphasis see Vickers, op. cit., pp. 93–116.

That is, rhetoric is guided by, or exploits, the image making capacity of the mind, just as logic does its ratiocinative capacity. As logic presents arguments in such a form that reason gives its assent to the conclusions, so rhetoric presents the conclusions of the author's reason in vivid images, whose persuasive force produces assent to the proposition, and movement towards the desired goal.[1] And although this places emphasis on the *artificial* character of rhetoric, Bacon insists that it should be used responsibly:

> The end of rhetoric is to fill the imagination with observations and images, to second reason, and not to oppress it. [IV, 456]

The 'Colours of Good and Evil'

Bacon's only two additions to rhetoric, or rather to the appendices to rhetoric (the promptuary store of raw material for rhetorical composition), both derive from his view of rhetoric as essentially 'insinuative'. The *Colours of Good and Evil* and the *Antitheses of Things* both contribute to the task of *colouring* an argument, that is, biasing it by piling together illustrations and observations which support a chosen aspect of the subject under consideration. These are *figures of thought* for embellishing and sustaining an argument, as opposed to the *figures of words* whose direct impression of the senses prompts assent by their pleasurable stimulus.

Bacon's *Colours of Good and Evil* is a fragment of a collection of commonplace generalisations to be used in *deliberatives*, that is, in discourse aimed at convincing an audience that a given course of action is desirable, or acceptable. This is done by showing in popular terms 'what is good and what is evil, and of good what is greater, and of evil what is the less' [VII, 77]. Each *colour* is a general precept which contributes to such discussions.[2] The

[1] Bacon's definition of rhetoric was evidently regarded as worthy of particular attention by readers of the period. The passage above appears as an entry in a commonplace book in Magdalene College, Cambridge. See W. T. Costello, *The Scholastic Curriculum at Early Seventeenth Century Cambridge* (Harvard, 1958), p. 55. Costello does not identify the passage as Bacon's.

[2] An early jotting related to this undertaking appears in Harl. MSS. 7017.16, fo. 46 (printed by Spedding [VII, 67–8]). This is headed 'Semblances or popularities of good and evill, with their redargutions; for Deliberacions'. It contains three groups of roughly noted 'colours' for making the best of an argument: 'Media via nulla est quae nec amicos parit, nec inimicos tollit', 'Ex malis moribus bonae leges', etc. This list does not include any of the colours from Bacon's published treatise, but those in the list are clearly intended for

colour is supported by illustrations, supporting comparisons, quotations and examples. Each colour is countered by an *elenchus*, which contradicts the original colour (but is equally a widely accepted commonplace), and supports its opposition by a comparable collection of counter-illustrations. This is best shown by an example:[1]

From something to nothing appears a greater step than from more to less; and again from nothing to something appears a greater step than from less to more. [IV, 465]

This colour is supported by, amongst others, the following observations:

It is a rule in mathematics that there is no proportion between nothing and something...Thus the loss of an eye is harder for a man with only one eye than for a man with two...Hence also the Sibyl, when she had burned her two first books, doubled the price of the third; for the loss of this would have been a degree of privation, and not of diminution...

With regard to the second part of this Sophism, it rests on the same foundation as the former...Hence the making of a beginning of anything is thought so great a matter;

Dimidium facti, qui bene coepit, habet, &c.

Hence also the superstition of astrologers, who make a judgment of the disposition and fortune of a man from the point or moment of his nativity or conception. [IV, 465–6]

Whilst amongst the illustrations which counter the colour are the following:

In the wasting of fortunes, the degree of debt which makes the first inroad on the capital seems worse than the last which reduces to beggary. To this belong the common forms; 'Sparing comes too late when all is gone;' 'as good never a whit as never the better,' &c. It deceives secondly, in respect of that principle of nature, that the decay of one thing is the generation of another; so that the degree of extreme privation is sometimes of less disadvantage, because it gives a handle and stimulus to some new course...in some cases the first beginnings of things are no more than what Epicurus in his philosophy calls *tentamenta*, that is imperfect offers and essays, which are nothing unless they be repeated or proceeded with. Therefore in this case the second

the same sort of occasion, and show once again Bacon's habit of noting down in groups fragments of material for the various sections of his 'promptuary'.

In the first edition of the *Essays* (1597), which includes the *Colours of Good and Evil*, these appear as *Places of perswasion and disswasion* in the title [VI, 521].

[1] The *Colours* reappear in the *De Augmentis* as an illustration of the deficiencies of rhetoric. The example here is taken from the treatment in the *De Augmentis*.

degree seems more worthy and more powerful than the first, as the wheel-horse in a cart does more work than the leader. Again, it is not a bad saying 'that it is the second word which makes the fray.' For perhaps the first would have passed. [IV, 466–7]

The colours themselves are sufficiently broad in application to be used in almost any context in which it is required to show that one course of action is preferable to another (because 'good' or 'less evil'), or that one course of action has more or less important consequences. The refutation of the colour serves either to counter an argument presented by an adversary, or equally, to provide an alternative course of argument. The illustrations of the colour and its refutation exploit all the supporting 'figures of thought' discussed in the last chapter, to persuade the listener by 'colours, popularities and circumstances', rather than by 'true and solid reasons' [VII, 77]. The colour is rendered more vivid and convincing in a particular context by piling up illustrative examples whose force is not that of rational evidence but of 'insinuative reason'. By neat juxtaposition of images, and the pleasing aptness of proverbs and apophthegms, conclusions are imposed on the reader without recourse to facts and formal reasoning.

The source for Bacon's *Colours of Good and Evil* is, on his own admission, Aristotle's discussion of the same topic in his *Rhetoric*.[1] The deliberative orator 'will never admit that he is recommending what is inexpedient or is dissuading from what is useful';[2] and since 'the expedient is good, we must first grasp the elementary notions of good and expedient in general'.[3] Aristotle proceeds to discuss a number of popularly accepted 'goods'. Any one of these, if appealed to in argument, will suffice to procure the assent of the audience. That is, if it can be shown by skilful argument that the course of action advocated leads to a popular 'good', the audience will be willing to accept the speaker's proposals.[4]

[1] *Rhetoric* I, VI and VII.
[2] *Rhetoric* I, III, 6.
[3] *Rhetoric* I, VI, 1.
[4] Quintilian gives more detailed instructions for weighting a case in one's favour by astute manipulation of the subjective elements of the case: motives, circumstances, attitudes, and so on. See particularly *Institutio Oratoria* 4, ii, 88–101. The Loeb edition has the following footnote on *colour*: '*color* is a technical term for the particular aspect given to a case by the skilful manipulator of the facts – the "gloss" or "varnish" put on them by the accused or accuser' (Peterson on Quintilian 10, i, 116). Whilst this biasing is once again related to Bacon's technique, Quintilian does not advocate the sort of imaginative weighting at

Bacon's appraisal of Aristotle's 'colours' indicates the difference between the two approaches:

First therefore I do not find the wisdom and diligence of Aristotle well pursued and supplied. For he began to make a collection of *the popular signs* or *colours of apparent good and evil,* both simple, and comparative; which are really the sophisms of rhetoric. Now these are of excellent use, especially for business and the wisdom of private discourse. But the labours of Aristotle regarding these colours are in three points defective; one, that he recounts a few only out of many; another, that he does not add the answers to them; and the third, that he seems to have conceived but a part of the use of them. For their use is not more for probation [supporting] than for affecting and moving. For there are many forms which, though they mean the same, yet affect differently; as the difference is great in the piercing of that which is sharp and that which is flat, though the strength of percussion be the same. [IV, 458]

The third point of difference is the crucial one. Aristotle's account contains such observations as the following:

In the case of doubtful goods, the arguments in their favour are drawn from the following. That is good the opposite of which is evil or the opposite of which is advantageous to our enemies.[1]
And things which produce a greater good are greater...And similarly, that which is produced by a greater cause; for if that which produces health is more desirable than that which produces pleasure and a greater good, then health is a greater good than pleasure.[2]

These are 'colours' exactly as Bacon uses the term – popularly accepted judgments of relative worth. But Aristotle makes no mention of supporting illustration and its independent power to persuade by making more 'piercing' the argument for pursuing a particular policy. Bacon advocates the use of all the figures of

which Bacon aims. Cicero uses *color* in a sense much closer to 'style' – the particular type of expression which characterises a passage. 'Ornatur igitur oratio genere primum et quasi colore quodam et suco suo' (*De Oratore* 3, 25 (95)). Wilson follows Cicero's usage, and applies the term to that group of figures which in other manuals appear as 'amplification', 'rhetorical sentences', or 'sententious figures': 'The third [type of figure] is when by diuersity of inuencion, a sentence is many wayes spoken, and also matters are amplified by heapynge examples, by dilatynge argumentes, by comparynge of thynges together, by similitudes, by contraries, and by diuers other lyke, called by Tullie Exornacion of sentences, or colours of Rhetorique' (*Arte of Rhetorique,* 1560 ed., fo. 87r). The colours, for Wilson, include 'Restyng upon a pointe' (fo. 90v), 'An euident, or plain settyng forthe of a thing as though it were presently doen' (fo. 91r), 'Askyng other, and answeryng our self' (fo. 93v), 'Snappishe askyng' (fo. 94r), similitude (fo. 96r), example (fo. 97r) and fable (fo. 100v).

[1] *Rhetoric* I, VI, 18.

[2] *Rhetoric* I, VII, 7.

'Exornation of Sentences' which bias and 'colour' by vivid imagery and by 'heapynge examples, by dilatynge argumentes, by comparynge of thynges together, by similitudes, by contraries, and by diuers other lyke',[1] to give added support to what is put forward as good or expedient. The strength of Bacon's technique is seen in the following series of illustrations to back up the refutation of the colour: *That which approaches to good or evil, is itself good or evil; but that which is remote from good is evil, that from evil, good*:

> You will never find flourishing underwood near great trees. And rightly was it said 'that the servants of a rich man are the greatest slaves.' So also the lower order of courtiers were pleasantly compared to the vigils of festivals, that are next the feast days, but are themselves devoted to fasting...evil also (especially in civil matters) betakes itself to good for concealment and protection. And hence malefactors seek the protection of sanctuaries, and vice itself resorts to the shadow of virtue;
>
> Saepe latet vitium promitate boni.
>
> So on the other hand good draws near to evil, not for company, but to convert and reform it. And therefore physicians attend more on the sick than the healthy; and it was objected to our Saviour that he conversed with publicans and sinners. [IV, 461]

Commonplace examples of poverty in the shadow of greatness, a witticism based on the proximity of penance and celebration, truisms about the tendency of evil to hide under the skirts of goodness, and a biblical allusion traditionally used to point out that goodness and humility seeks out vice, to convert it, all appear to bear out the fact that the colour (despite its truistic appeal) is ill-founded. In fact, the support is illusory. The match of the flourishing vegetation, the rich, the well fed, and the healthy with the 'good' is accepted as telling for the argument by insinuation. The aptness of many of the examples is accepted on the strength of their familiarity, and the cumulative effect of successive illustrations taken as providing overwhelming justification.

Bacon's systematic use of evocative illustration to strengthen the appeal of a colour, or to reduce the appeal of an opponent's colour, greatly extends the scope of such sophistical means of supporting a case. By careful choice of illustration the speaker can ensure that the colour is presented in terms particularly

[1] Wilson, *Arte of Rhetorique*, 1560 ed., fo. 87r.

suited to audience and occasion. By introducing a carefully selected counter-example he may succeed in making his opponent's position appear untenable.[1]

'Antitheses of Things'

For Bacon, equivocations which influence the audience through a carefully planned strategy of image building and illustrative exemplification are the backbone of rhetorical argument. The *Antitheses of Things* is a collection of observations in a similar vein to those used in support of, or opposition to selected colours. The *antitheses* are appealing, epigrammatical statements for and against such commonplace topics as 'truth', 'nobility', 'fortune', presented not for their intrinsic truth or falsehood, but for their persuasive force under particular circumstances.[2] Thus under the heading 'loquacity', Bacon gives the following propositions *for* the title topic:

> He that is silent betrays want of confidence either in others or in himself.
> All kinds of constraint are unhappy, that of silence is the most miserable of all.
> Silence, like night, is convenient for treacheries.
> Thoughts are wholesomest when they are like running waters.
> [IV, 484–5]

He gives the following *against* the topic:

> Silence gives to words both grace and authority.
> Silence is the sleep which nourisheth wisdom.
> Silence is the fermentation of thought.
> Silence is the style of wisdom.
> Silence aspires after truth. [IV, 484–5]

Once again, such observations are to be used as illustrative supports for particular stands taken on some subject (like 'lo-

[1] Weighted examples were used in a less subtle way to make partisan points in polemical religious writings in the period. For the way in which weighted illustration was used in this context see R. Pineas, 'Polemical *exemplum* in sixteenth century religious controversy', *Bibliothèque d'Humanisme et Renaissance*, 28 (1966), pp. 393–6.

[2] These are the sort of headings used by any student in his commonplace book as a guide to his collection of striking sentences from ancient and modern authors. See C. Hoole, *A new discovery of the old arte of teaching school* (London, 1660); R. R. Bolgar, *The Classical Heritage and its Beneficiaries* (Cambridge, 1958), pp. 272–4 and passim. See also any sixteenth-century commonplace book.

quacity') in the course of an extended discussion. They provide an armoury of pithy and persuasive arguments for all occasions. Like the *Colours of Good and Evil* they do this by specifying in each case a vivid context for the attribute in question. Each proposition provides an angle from which the subject under discussion may be viewed in a particularly revealing light. The theme can then be elaborated upon as the author desires:

I would have in short all topics which there is frequent occasion to handle (whether they relate to proofs and refutations, or to persuasions and dissuasions, or to praise and blame)[1] studied and prepared beforehand; and not only so, but the case exaggerated both ways with the utmost force of wit, and urged unfairly, as it were, and quite beyond the truth. And the best way of making such a collection, with a view to use as well as brevity, would be to contract those commonplaces into certain acute and concise sentences; to be as skeins or bottoms of thread which may be unwinded at large when they are wanted. [IV, 472]

Such *sententiae* were an integral part of the renaissance rhetorician's equipment.[2] Bacon's *Antitheses of Things* systematises the extremely opportunistic use to be made of these. The truth or falsehood of the *sententiae* is not assessed, and is irrelevant. They stand in compact form for attitudes towards a question, which may be appealed to wherever the discourse calls for a particular supporting angle.[3] If this appears to us an extremely artificial way of writing or speaking creatively, this only goes to emphasise

[1] These are the three traditional types of oration. See for instance, Cicero, *De Inventione* I, 7. The three types are judicial (proofs and refutations), deliberative (persuasions), and epideictic (praise or blame).

[2] Collections of *sententiae* abound in the period. The reader of sixteenth-century authors who is dazzled by the wealth of *sententiae* quoted from the classics should bear in mind that there was an industry of works with titles like the following (bound together), selected at random: *Marci Tullii Ciceronis, Sententiae illustriores, Apophthegmata item, et Parabolae siue similia: aliquot praeterea eiusdem piae sententiae. Authore Petro Lagnerio Compendiensi* (1546); *Sententiae et Proverbia ex Plauto, Terentio, Virgilio, Ouidio, Horatio, Iuuenale, Persio, Lucano, Seneca, Lucretio, Martiale, Sillio Italico, Statio, V. Flacco, Catullo, Propertio, Tibullo, Claudiano* (1511); *Sententiae Singvlis versibus contentae, iuxta ordinem literarum, ex diuersis poetis* (1540). The first and third of these list the *sententiae* under topics, including some of those used by Bacon. I have not come across an author who sets out *sententiae pro* and *con* a given theme, apart from Bacon.

[3] Harl. MSS. 7017.16, fo. 26r, contains these 'antitheses of things': 'Vpon quaestion to reward evill with evill', 'vpon quaestion whether a man should speak or forbear speach' [VII, 209]. These rough notes show one essential characteristic of the antitheses: their format is designed to allow the individual to add to his *pro* and *con* lists under any heading as he comes across appropriate material in his reading. The *sententiae* which Bacon has listed under the three headings have clearly been added at various times (the inks and direction of slope of the handwriting vary from entry to entry).

how deeply Bacon is committed to a rhetoric in which every move in a discourse is planned so as to insinuate the desired conclusion into the mind of the audience.[1]

[1] Harl. MSS. 7017.16 in its entirety gives ample evidence of Bacon's dedication to self-conscious and contrived rhetorical expression. The manuscript contains page after page of 'prefaces, conclusions, digressions, transitions, intimations of what is coming, excusations' [IV, 492], as well as proverbs in several languages, *sententiae,* notes on sources of comparison (e.g. 'Hercules pillers non vltra' (fo. 3r); 'Charons fare' (fo. 18v)), apophthegms and so on. It is also striking how many of the jottings find their way into Bacon's subsequent works.

13

The method of Bacon's essays

Bacon's *Essays*[1] fall squarely under the heading of presentation or 'method of discourse'.[2] They are didactic, in Agricola's sense of presenting existing knowledge to someone in a form in which it may be believed and assimilated. Whatever methods they employ must therefore be 'magistral' as opposed to 'initiative' (see above, chapter 9). Bacon does not set out to give formal justification for particular social and political beliefs. All social and political precepts are ultimately matters of opinion, and these do not lend themselves to demonstrative or aphoristic presentation. ('For the method of the Stoics, who thought to thrust virtue upon men by concise and sharp maxims and conclusions, which have little sympathy with the imagination and will of man, has been justly ridiculed by Cicero' [IV, 456].) In order to appear convincing some more 'insinuative' method of presentation is required. On the other hand, the tone of the essays precludes their being regarded as amusing exercises in

1 The studies of Bacon's *Essays* are practically endless. Amongst those who have cast interesting light on the *Essays* from a variety of points of view are the following: M. A. Bowman, *The English Prose Style of Sir Francis Bacon* (unpublished Ph.D. diss., Wisconsin, 1964); B. W. Vickers, *Francis Bacon and Renaissance Prose* (Cambridge, 1968); P. B. Burke, 'Rhetorical considerations of Bacon's style', *College Composition and Communication*, 18 (1967), pp. 23–31. Bowman devotes almost half her thesis to a minute stylistic analysis of all Bacon's essays, with careful collation of editions; Vickers discusses the revisions in the *Essays* and their stylistic consequences (pp. 217–31); Burke gives a brief, neat characterisation of the modifications of successive editions of the *Essays* as clarifications of particular points of view. I have not seen S. Mandeville, *The Rhetorical Tradition of the Sententia; with a Study of its Influence on the Prose of Sir Francis Bacon and of Sir Thomas Browne* (unpublished Ph.D. diss., St. Louis, 1960), but the abstract is intriguing.

2 The fact that the essays use rhetorical devices at all precludes their contributing directly to the natural histories on which the inductive method is based (contrary to the claims of some authors, notably R. S. Crane), since Bacon specifically excluded the ornamental and anecdotal from natural history. See Crane, 'The relation of Bacon's *Essays* to his programme for the advancement of learning', *Schelling Anniversary Papers* (New York, 1923), pp. 87–105.

rhetorical equivocation. The *Essays* are, I shall suggest, carefully constructed to put across practical precepts which Bacon believed to be of value to men of all intellectual backgrounds. The essay form is used as a 'method' for projecting these precepts in an appealing and readily acceptable form.[1]

The essay 'Of Ceremonies and Respects' [VI, 500–1] opens with two *sententiae*, one based on comparison through a proverb, and an apophthegm:[2]

> He that is only real, had need have exceeding great parts of virtue; as the stone had need to be rich that is set without foil. But if a man mark it well, it is in praise and commendation of men as it is in gettings and gains: for the proverb is true, *That light gains make heavy purses*; for light gains come thick, whereas great come but now and then. So it is true that small matters win great commendation, because they are continually in use and in note: whereas the occasion of any great virtue cometh but on festivals. Therefore it doth much add to a man's reputation, and is (as queen Isabella said) *like perpetual letters commendatory*, to have good forms. [VI, 500]

These remarks are laid out as if they formed a reasoned sequence: statement, subsidiary statement ('But...'), conclusion ('Therefore...'). In fact the progression is not reasoned but suggested. Two general moral precepts are underlined by pre-

[1] I have chosen to look at the 1625 *Essays*, as showing Bacon's skills in presentation at their most mature. Several critics have shown that there is continuous development from the original 1597 edition of ten essays to the final 1625 edition of fifty-eight essays. E. Arber, *A Harmony of the Essays...of Francis Bacon* (London, 1871) prints the successive editions in parallel columns, which makes comparison convenient. The modifications of early essays in later editions always take the form of expansion of the original, frequently at the expense of the stylistic balance of clauses and the overall compactness of the essay. See Vickers, *Francis Bacon and Renaissance Prose* (Cambridge, 1968), pp. 217–31; Bowman, op. cit.

[2] I have tried to consider a representative cross-section of the different types of essay, and to show that the tactics of presentation are the same in each. I have used Bowman's classification of the essays as a whole into stylistic groups as the basis for my own selection. She groups the 1597 essays (which give a basis for consideration of the types of later essay) as follows: '"Of Studies," "Of Ceremonies and Respects," "Of Expence," and "Of Regiment of Health" are all similar in the sharp, antithetical shape of their sentences. In these essays balance and opposition of thought are suggested by such structural signals as imperatives and opposition of conjunctives. "Of Discourse," "Of Sutors," "Of Faction," and "Of Negociating" constitute a contrasting group of essays by the preceptive weight of individual sentences, integrated rather by smaller syntactic clues than by any larger structural organization. In these latter pieces logical force gives way to simple power as observation or precept sums up experience and indicates practice. Finally, in a curious way "Of Honour and Reputation" and "Of Followers and Friends" respectively display both the quality of logical integration and of preceptive looseness even while showing the new tendency of ordering by definition and classification' (pp. 197–8). I have chosen to consider 'Of Ceremonies and Respects' from the first of these groups, 'Of Suitors' from the second, and 'Of Followers and Friends' from the third.

sentation in the form of familiar proverbial sayings: the man who can afford to behave quite naturally must be exceptionally virtuous (as only a flawless jewel can do without artificial high-lighting of its good points);[1] repeated small occasions for moderate praise produce a greater overall effect of merit than isolated occasions for loud acclaim (as small savings accumulate into wealth, and everyday occurrences outnumber festivals). The title of the essay specifies a context for these two observations: only an exceptional man can dispense with the public conventions of virtue (i.e. with 'ceremonies and respects'); repeated small observances of the details of decorum are worth more than isolated acts of greater significance (virtuous acts in this context rank lower than acts of decorum).[2] Here it is deliberately assumed that praise is a commodity to be bargained for: 'it is in praise and commendation of men as it is in gettings and gains'.[3] This attitude is reiterated in Queen Isabella's apophthegm: 'good forms' should always be observed, just as it is always prudent to carry good references.[4] There is no formal relation between these three sentences, but together they provide a setting for the discussion which follows. Propriety is the public counterpart of virtue; discreet moderation is effective in manners as in more weighty actions.

In the 1597 edition of the *Essays* this opening is separated from the rest of the essay in a paragraph of its own. In the later edition the paragraphing is implied by a change in tone. The opening is followed by a collection of disparate observations and precepts on decorum. Whereas in the introduction good forms are discussed as a foil for inner virtue, the later discussion is strictly concerned with strategic use of manners for personal advancement. There has therefore been a considerable narrowing of interest in the direction of social expediency, taking up the theme of 'best dividend on capital invested' from the introduction. In this context the burden of the discussion is the

1 This comparison is used in another context in the 1612 essay 'Of Beauty': 'Vertue is like a rich stone, best plain set: and surely vertue is best set in a body that is comely though not of delicate features' [VI, 569]. See Harl. MSS. 7017.16, fo. 2r: 'Vertue like a rych gemme [?] best plaine sette.'

2 In the Latin translation 'decorum' occurs in the essay title: 'De Caeremoniis Civilibus, et Decoro' [VI, 500, note 3].

3 In the 1597 version this phrase runs: 'it is in praise as it is in gains'. The expansion underlines the commercial attitude.

4 Isabella's apophthegm appears in Bacon's published apophthegms [VII, 139].

advisable limitations to be placed on 'good forms' for the greatest social profit:

To attain [good forms] it almost sufficeth not to despise them; for so shall a man observe them in others; and let him trust himself with the rest. For if he labour too much to express them, he shall lose their grace; which is to be natural and unaffected. Some men's behaviour is like a verse, wherein every syllable is measured; how can a man comprehend great matters, that breaketh his mind too much to small observations? Not to use ceremonies at all, is to teach others not to use them again; and so diminisheth respect to himself; especially they be not to be omitted to strangers and formal natures; but the dwelling upon them, and exalting them above the moon, is not only tedious, but doth diminish the faith and credit of him that speaks. And certainly there is a kind of conveying of effectual and imprinting passages amongst compliments, which is of singular use, if a man can hit upon it.[1]

[VI, 500–1]

As this passage shows, the discrete observations all stress in various ways that 'good forms', whilst necessary, must remain the clothing of serious political purpose (although it is not suggested that such purposes should be morally laudable), and must not dominate to the neglect of business.[2] The precepts instructing the reader on the extent to which decorum should be observed are interspersed with sententious observations which depend upon the preceding precept (as the second and fourth sentences in the passage above show). The theme is illustrated in the closing sentences of the essay by a proverb and an apophthegm:

It is loss also in business to be too full of respects, or to be curious in observing times and opportunities. Salomon saith, *He that considereth the wind shall not sow, and he that looketh to the clouds shall not reap.* A wise man will make more opportunities than he finds. Men's behaviour should be like their apparel, not too straight or point device, but free for exercise or motion. [VI, 501]

This closing passage (the proverb added in 1612, the apophthegm in 1625) confirms the extremely restricted and pragmatic

[1] The Latin translation makes the sense of the last phrase clearer: 'modus artificiosae cujusdam insinuationis, in verbis ipsis, inter formulas communes, qui homines revera inescat et mirifice afficit.' Significant words can be interspersed between formal compliments in a particularly effective way.

[2] As Bowman shows, the emphasis on a balance to be observed between affectation in manners and neglect of good forms is supported by stylistic balance of clauses, op. cit., pp. 209–11.

context in which 'ceremonies and respects' have been confined in the essay, despite the misleading generality of the opening. The passage occurs in very similar form in the *Advancement of Learning* (1605) amongst the provisional observations which Bacon sets down as part of the 'wisdom of conservation', or that knowledge which will enable a man to conduct himself successfully in public affairs. The emphasis in this discussion is openly on outward show for personal advancement [III, 445–7]. In both contexts the proverb and apophthegm illustrate a concluding position on decorum, to give a particular view emphasis. As with the opening of the essay, these closing illustrations are effective by juxtaposition. Although they are given the status of a conclusion, as if they had arisen as the consequences of the preceding discussion, there is in fact no logical link.[1]

The addition of proverb and apophthegm in 1612 and 1625 provides a concluding emphasis for the main point of the essay. The other minor additions also help to give the essay a clearer focus. Bacon appears to have felt the need for further support for his persuasive points.[2]

The essay 'Of Suitors' also appeared first in the editions of 1597. It opens with a section (a paragraph in the original edition) summarising the wrong motives which can lead one individual to take up another's cause. The opening sentence is modified in an interesting way in the course of the various editions. In its original form its 'almost abstruse ellipsis'[3] simply establishes a tone of disapproval:

Manie ill matters are vndertaken, and many good matters with ill mindes. [VI, 528]

[1] The similarity between the passage in the *Advancement of Learning* and the essay points to the fact that Book 2 of the *Advancement* is an 'essay' in the classification of knowledge (as Book 1 is an 'essay' in the purpose, and abuse of learning). Bacon maps out convincingly a plan for the reorganisation and expansion of the curriculum subjects. In indicating the *desiderata* and in sketching the form contributions to neglected areas should take he is above all concerned to convince his audience of the need for such studies. The specimens do not therefore illustrate formal techniques, but indicate the possibility of systematic study in these areas.

[2] Also missing from the 1597 essay are Isabella's apophthegm (it is 'like perpetual letters commendatory' to have good forms); a balancing clause for the recommendation not to omit good forms altogether ('but the dwelling upon them and exalting them above the moon, is not only tedious, but doth diminish the faith and credit of him that speaks'); the next observation, that care with forms of words can insinuate one's point of view; and the cautionary sentence beginning: 'Men had need beware how they be too perfect in compliments'. The apophthegm completes the illusion of a reasoned opening, as discussed above. The further additions confirm the moderate view that a profitable mean must be preserved between affectation and lack of decorum.

[3] Bowman, op. cit., p. 223.

In the 1625 edition this opening is expanded into two sentences:

> Many ill matters and projects are undertaken; and private suits do putrefy the public good. Many good matters are undertaken by bad minds; I mean not only corrupt minds, but crafty minds, that intend no performance. [VI, 495]

The expansions modify rather than clarify the sense of the original. Two new ideas are introduced: suits undertaken for private benefit 'putrefy the public good', and hence fall under 'ill matters'; good suits may be abused not only by those 'bad minds' which consciously distort them, but equally by those who plan to use suits for their own purposes, and then discard them. The first addition provides a subsidiary theme which is taken up again in the closing sentence of the essay (also an addition to the 1597 version):

> There are no worse instruments than these general contrivers of suits; for they are but a kind of poison and infection to public proceedings. [VI, 497]

The second addition to the opening provides a transition to the catalogue of wrong motives (private or personal) for supporting suits, which is followed by a short section of precepts on the theme that all suits once made are in some sense justifiable, either by a natural right, or in the eye of the judge.

Two positions are established as the result of the 'colouring' of the opening. Support of private suits corrupts the business of the state; any suit whatsoever may be judged worthy of support, once it has been undertaken from public rather than private motives. We thus have the position that suits undertaken in the public interest may be successfully pressed, according to Bacon. From this viewpoint, the bulk of the essay consists of a collection of miscellaneous precepts giving practical advice on the effective pressing of suits so as to balance the interests involved, on the understanding that this advice is given in the context of the tactical pursuit of public ends through suits. The opening and close provide a setting which gives cohesion to the essay as a whole (leading up to the improvement of the 1625 version of the essay, in which the key sentences in these are added). Once it is accepted that suits and suitors are to be considered as political instruments, the theme of tactical use is accepted for the precepts as a whole.

'Of Followers and Friends' is the last of the essays first published in 1597 which I shall consider. Here once again, the opening clearly defines the scope of the discussion:

> Costly followers are not to be liked; lest while a man maketh his train longer, he make his wings shorter. I reckon to be costly, not them alone which charge the purse, but which are wearisome and importune in suits. Ordinary followers ought to challenge no higher conditions than countenance, recommendation, and protection from wrongs.
> [VI, 494]

The point under discussion is the desirable balance between the prestige of followers, who increase a man's reputation by their number and behaviour, and expenditure both of money and time (which may reduce prestige in the end). Once again the emphasis is on use for personal promotion in public affairs. In the subsequent discussion the problem of balance is shifted to that between the aspirations of the follower and those of the man he follows. Followers who intend to use their patron for their own purposes only are to be avoided. The 'most honourable kind of following' is where follower and followed gain equally from the alliance, the patron gaining in glory, the follower in support. In this context mutual dependence is presented as a strategic balancing of indebtedness; the patron should not give too much to too many, but should take advantage of the honest advice to be elicited from followers under the right conditions. This theme is reinforced in conclusion by an apophthegm, and a coda on friendship:

> To take advice of some few friends is ever honourable; *for lookers-on many times see more than gamesters; and the vale best discovereth the hill.* There is little friendship in the world, and least of all between equals, which was wont to be magnified. That that is, is between superior and inferior, whose fortunes may comprehend the one the other. [VI, 495]

The shift to 'friendship' from 'followers', like the pairing of the terms in the title, is justified only if mutual self-interest is the motive for the one as for the other. Politically, a disinterested companion close enough to tell the truth is a sound investment. And this is the only side of friendship with which Bacon is here concerned, or which can be discussed in the narrow context dictated by the carefully coloured opening. This worldly assessment of friendship provides a striking and appropriate close to a lesson in the exploitation of dependents. It should be noted that the view of friendship projected as a rhetorical flourish in this essay is at odds with the sensitive and optimistic treatment of the same subject in the essay 'Of Friendship', where it is discussed in a different context.

A single principle of organisation is used in the three stylistic groups of the early essays, and this is sustained and supported in the versions of these essays in the later editions. Basically these essays communicate precepts for the guidance of personal conduct in public affairs, based on Bacon's own political experience. The context for these precepts is extremely carefully specified by the opening of the essay, so that the broad generality of the observations is effectively narrowed into some particular set of circumstances, in which it may be interpreted in a particularly useful way. Attention is focused on a particular aspect of the general topic specified by the essay title, by means of careful, methodical colouring of the type which Bacon regarded as an essential part of the 'wisdom of transmission'.

This reading characterises the 1597 essays as a condensation of practical precepts on chosen topics relating to business and policy, backed up by the maximum compression of colouring, persuasive and illustrative *exempla*, and interspersed with apt apophthegms. These essays are (and remain in their later versions) straightforwardly instructive, but without any attempt to justify rationally the knowledge which they transmit. The precepts are retailed as proven from experience, and backed up by any material which will contribute to their acceptance by the reader. The *derivation* of such precepts is, it goes without saying, a distinct and unrelated activity. It involves careful observation of men in society, and of past political events and their outcomes.

The knowledge presented in the essays discussed above, and in the other essays dating from the 1597 edition, fills a gap in contemporary instruction, according to Bacon's own account in the *De Augmentis:*

The science of negociation has not hitherto been handled in proportion to the importance of the subject, to the great derogation of learning and the professors thereof. For from this root springs chiefly that evil, with which the learned have been branded; '*That there is no great concurrence between learning and practical wisdom.*' For if it be rightly observed, of the three wisdoms which we have set down to pertain to civil life, the wisdom of behaviour is by learned men for the most part despised, as a thing servile, and moreover an enemy to meditation. For wisdom of government, it is true that as often as learned men are called to the helm, they acquit themselves well, but that happens to few. But for the wisdom of business (of which I am now speaking), wherein man's life is most conversant, there are no books at all written of it, except some few civil advertisements collected in one or two little volumes, which have no proportion to the magnitude of the subject. [V, 35]

Whence it appears that there is a wisdom of counsel and advice even in private causes, arising out of a universal insight and experience of the affairs of the world; which is used indeed upon particular causes, but is gathered by general observation of causes of like nature. [V, 36]

In the early essays, I suggest, Bacon passes on his own 'universal insight and experience' in a form palatable to a general audience, and one which makes easy the quick retrieval of general precepts for application in particular cases. These are, for Bacon, the essential attributes of a method of presentation which fulfils the requirements of successful transmission of knowledge. Such essays thus form part of the *Doctrine concerning Scattered Occasions* [V, 35], which in Bacon's view will enable every individual to run his own life profitably and efficiently. 'There [is] nothing in practice, whereof there is no theory and doctrine' [V, 59], according to Bacon; even the individual who is not equipped to discover the theory himself can be persuaded of the truth of the corresponding practical rules, for his personal benefit.

The essay 'Of Riches' was first introduced in the 1612 edition. Its opening straight away establishes the context in which Bacon intends to discuss the subject: wealth is for *use.* Wealth for its own sake is of no conceivable value, and can be a positive inconvenience:

I cannot call Riches better than the baggage of virtue. The Roman word is better, *impedimenta.* For as the baggage is to an army, so is riches to virtue. It cannot be spared nor left behind, but it hindereth the march; yea and the care of it sometimes loseth or disturbeth the victory. Of great riches there is no real use, except it be in the distribution; the rest is but conceit. [VI, 460]

The judgment that riches are an 'impediment' is supported by a comparison based on a verbal play on 'impedimenta'. 'Impedimenta' means in Latin both a hindrance, and the baggage and supplies of an army; therefore riches (which in Bacon's judgment are a hindrance) are like the baggage of an army, a hindrance, but necessary for the action. Moreover, like the army's baggage, riches may in themselves become a preoccupation which distracts the individual from his pursuit of worthy ends in favour of rescuing his riches. Nothing, of course, has been proved by this argument. But the extended play on 'impedimenta', with its close matching of salient details, fixes in the reader's mind an image of riches as a tangible, bulky benefit, which provides its own hazards, but sustains all action.

To support the claim that riches in themselves give no particular power to a man Bacon continues as follows:

> Do you not see what feigned prices are set upon little stones and rarities? and what works of ostentation are undertaken, because there might seem to be some use of great riches? But then you will say, they may be of use to buy men out of dangers or troubles. As Salomon saith, *Riches are as a strong hold, in the imagination of the rich man.* But this is excellently expressed, that it is in imagination, and not always in fact. For certainly great riches have sold more men than they have bought out. [VI, 460]

Riches are of 'no solid use to the owner'; their only power is to purchase small and useless prestige articles. Moreover, although the rich imagine themselves to be made invulnerable by their wealth, their 'fortress' is a figment of their imagination. More men sell themselves for riches than are saved (bought out of danger) by the outlay of riches. In this case the use of the proverb is two-edged: at first sight it appears to support the suggestion (put in the mouth of the reader) that riches protect the rich man, but on closer inspection Solomon's observation turns out to support Bacon's opposed claim that no *real* protection is provided, apart from the psychological sense of invulnerability. And read in this way the proverb provides a jumping-off ground for the damaging observation, which closes the opening section (unchanged in the 1625 version), that 'great riches have sold more men than they have bought out'.

The body of the essay collects together precepts for the instruction of those who wish to become rich. These are interspersed with comment and illustration emphasising the dubious dealings, and fraudulent means usually necessary for amassing wealth. The tone of disapproval is sustained, but the advice is reasonable, and the observations accurate. Most of this material appears for the first time in the 1625 version of the essay; the 1612 version moves almost immediately from the opening section persuading of the uselessness of wealth to a closing section (which in unchanged form closes the 1625 version) warning against vain hoarding of wealth, since unless it is freely dispensed during the owner's lifetime it will only be squandered and lost in the future:

Be not penny-wise; riches have wings, and sometimes they fly away of themselves, sometimes they must be set flying to bring in more. Men leave their riches either to their kindred, or to the public; and moderate portions prosper best in both. A great state left to an heir, is as a lure to all the birds of prey round about to seize on him, if he be not the better stablished in years and judgment. Likewise glorious gifts and foundations are like *sacrifices without salt*;[1] and but the painted sepulchres of alms, which soon will putrefy and corrupt inwardly. [VI, 462]

Wealth, whilst it is necessary, is only a good thing (and a spur to virtue) if it is kept in circulation, and made to work for its owner and society. Mere accumulation of riches adds nothing to a man's real stature, or to his security, nor does it assure the well-being of his descendants.

Each point in this clear statement of the real value of wealth is made by careful colouring, illustration and exemplification. By piling up illustrations Bacon conveys his judgment without either justifying his attitude in rational terms, or openly stating his bias. He therefore succeeds in communicating a body of instruction on the accumulation of riches, and the use to be made of them, which is in fact placed strictly in the context of firm ethical condemnation of the miser and of conspicuous flaunting of wealth for its own sake. The overall bias of the essay stands out most clearly if we compare the passages above with a passage from the *De Augmentis* in which Bacon states openly his views on riches:

In the second place [for advancing a man's fortune] I set down wealth and means, which many perhaps would have placed first, because of

[1] This phrase is an addition to the 1625 version of the essay – the only one in the carefully constructed opening and closing sections.

their great use in everything; but that opinion I may condemn, for the reason which Machiavelli gave in a case not much unlike. For whereas there was an old proverb, 'that money is the sinews of war,' yet he maintained on the contrary that the true sinews of war are nothing else than the sinews of a valiant and military people. And so in like manner it may be truly affirmed, that it is not money that is the sinews of fortune, but it is rather the sinews of the mind, wit, courage, audacity, resolution, temper, industry, and the like. [V, 72]

Assertions like: 'that opinion I may condemn', 'it may be truly affirmed', do not occur in the essays. Bacon presents his views persuasively, and as far as possible conclusively, but without openly insisting. The reader follows his points of his own volition; he turns away from consideration of wealth for its own sake without any open statement on Bacon's part that this attitude is to be abhorred.

'Of Simulation and Dissimulation' first appeared in the 1625 edition of the *Essays*. It opens, in a way with which we have by now become familiar, by specifying a context for the discussion:

Dissimulation is but a faint kind of policy or wisdom; for it asketh a strong wit and a strong heart to know when to tell truth, and to do it. Therefore it is the weaker sort of politics that are the great dissemblers. [VI, 387]

This establishes at the outset that feigning appearances is inferior to open dealings and truthfulness. This 'entrance' is immediately qualified: only the most perfect rulers are capable of acting entirely without dissembling. Bacon next makes the point that dissimulation is not the same as shrewdness in political dealings, and that dissimulation is merely a substitute policy where shrewdness is lacking. This he does by means of a short rhetorical induction, or sequence of examples supporting the point:

Tacitus saith, *Livia sorted well with the arts of her husband and dissimulation of her son*; attributing arts or policy to Augustus, and dissimulation to Tiberius. And again, when Mucianus encourageth Vespasian to take arms against Vitellius, he saith, *We rise not against the piercing judgment of Augustus, nor the extreme caution or closeness of Tiberius.* These properties, of arts or policy and dissimulation or closeness, are indeed habits and faculties several, and to be distinguished. [VI, 387]

This distinction, and the restricting of dissembling to the weaker sort of ruler, are further supported by comparison. Great rulers see clearly the outcomes of their policies; weaker men have less grasp of circumstances:

For where a man cannot choose or vary in particulars, there it is good to take the safest and wariest way in general [and dissemble]; like the going softly, by one that cannot well see. Certainly the ablest men that ever were have had all an openness and frankness of dealing; and a name of certainty and veracity; but then they were like horses well managed; for they could tell passing well when to stop or turn.
[VI, 387]

Shrewdness is the quality possessed by great rulers, whereby by careful assessment of any situation they 'tell passing well when to stop or turn', and control events by astute but honest dealing. Dissimulation is a second-rate substitute for shrewdness. We now have a context for detailed consideration of the conditions and varieties of dissimulation.

Bacon next distinguishes by explicit partition three levels of counterfeiting appearances in public:

There be three degrees of this hiding and veiling of a man's self. The first, Closeness, Reservation, and Secrecy; when a man leaveth himself without observation, or without hold to be taken, what he is. The second, Dissimulation, in the negative; when a man lets fall signs and arguments, that he is not that he is. And the third, Simulation, in the affirmative; when a man industriously and expressly feigns and pretends to be that he is not. [VI, 387–8]

It is a feature of the essays introduced for the first time in 1625 that their greater length is made manageable by the use of partition to announce the development of the theme to the reader in advance. Each of the divisions is then treated in turn, using comparison and exemplification to colour the treatment, so that the final judgment appears as the obvious outcome of the account. For instance, in the present example, under the division 'Secrecy' Bacon argues as follows:

But if a man be thought secret, it inviteth discovery; as the more close air sucketh in the more open; and as in confession the revealing is not for worldly use, but for the ease of a man's heart, so secret men come to the knowledge of many things in that kind; while men rather discharge their minds than impart their minds. In few words, mysteries are due to secrecy. Besides (to say truth) nakedness is uncomely, as well in mind as body; and it addeth no small reverence to men's manners and actions, if they be not altogether open. [VI, 398]

Remembering always (from the introductory stage-setting) that openness in the hands of the gifted ruler is superior to all feigning of appearances, simple secrecy is endorsed as a politic procedure in civil affairs. The support for this view derives largely from a collection of *sententiae* drawn from the lists *pro* 'Silence in matter of Secrecy' and 'Dissimulation' in the *Antitheses of Things* [IV, 483, 485]. (One of the *sententiae* used *against* 'Silence in matter of Secrecy' is used in favour of the same topic in the essay: 'Silence is the virtue of a confessor.')

Dissimulation is coloured so as to appear an inferior branch of secrecy, despite the inevitable necessity of using it under some political circumstances:

So that no man can be secret, except he give himself a little scope of dissimulation; which is, as it were, but the skirts or train of secrecy. [VI, 388–9]

Simulation as a regular procedure is condemned outright.

The final paragraph of the essay qualifies the disparagement of dissimulation, and the condemnation of simulation as vicious and impolitic, by listing political circumstances under which one or the other of these may be necessary. This qualified endorsement is immediately undermined by a statement of three practical drawbacks to habitual use of deception in politics:

There be also three disadvantages, to set it even. The first, that simulation and dissimulation commonly carry with them a shew of fearfulness, which in any business doth spoil the feathers of round flying up to the mark. The second, that it puzzleth and perplexeth the conceits of many, that perhaps would otherwise co-operate with him; and makes a man walk almost alone to his own ends. The third and greatest, is, that it depriveth a man of one of the most principal instruments for action; which is trust and belief. [VI, 389]

The carefully made distinctions, and their internal comparison, allow Bacon to explore the nuances of political counterfeiting, and to present at the end of the essay a general precept which lays down the scope of application of carefully graded degrees of dissimulation:

The best composition and temperature is to have openness in fame and opinion; secrecy in habit; dissimulation in seasonable use; and a power to feign, if there be no remedy. [VI, 389]

Because of the preoccupation with distinctions within the topic, the discussion focuses squarely on civil expediency, and avoids the question of whether dissembling is morally acceptable in politics.[1] Moral judgment, for Bacon, is not a useful part of practical politics.

The full title of the 1625 *Essays* is: *The Essayes or Counsels, Civill and Morall, of Francis Lo. Verulam, Viscount St. Alban* [VI, 371]. The essays which I have so far considered have been concerned with civics, as Bacon defined the subject. As the title suggests, some of the essays are concerned with moral questions, and these are presented in a slightly different manner.

The essay 'Of Envy' is an exploration of popular views on envy, and the circumstances under which people envy and are envied, and the way in which private, personal envy differs from public. The essay gives no explicit advice; it is a collection of observations on envy's nature and occurrence.

In the body of the essay Bacon uses the same techniques of comparison and illustration which were discussed in the preceding chapters, and which govern the development and organisation of the essays so far considered. The opening passage, however, consists of a discussion of the powers of envy and love to fascinate (to give the evil eye), which corresponds to a topic discussed in the *Sylva Sylvarum* [II, 653]. Having made the suggestion, which he evidently regards as a serious one, that the two affections may share certain physical powers, Bacon moves on to a partitioned discussion of occurrences of envy, supported by *sententiae*, and by antitheses from under the headings 'Nobility' and 'Envy' (one such sentence is transferred from the 1612 essay 'Of Nobility'). The following passage is characteristic:

A man that is busy and inquisitive is commonly envious. For to know much of other men's matters cannot be because all that ado may

[1] In his miscellaneous hints for swaying the opposition in *Topics* VIII, 2 Aristotle advocates the use of internal division of a topic both as an ornament to the discussion, and as a way of distracting attention from objections to the topic as a whole. His entire discussion of artificial devices to be used in argument has many points in common with Bacon's tactics in 'argument' in the essays.

concern his own estate; therefore it must needs be that he taketh a kind of play-pleasure in looking upon the fortunes of others. Neither can he that mindeth his own business find much matter for envy. For envy is a gadding passion, and walketh the streets, and doth not keep home: *Non est curiosus, quin idem sit malevolus.*

Men of noble birth are noted to be envious towards new men when they rise. For the distance is altered; and it is like a deceit of the eye, that when others come on they think themselves go back. [VI, 393]

The circumstances of envy are presented in every case in such a way as to denigrate the envier, whilst persuasively indicating the cause of the affection. In contrast to this entirely unfavourable, exploratory painting of private envy, the discussion of public envy returns to the expedient discussion of the civil essays. The only precepts to figure in the essay are concerned with the effects of public envy, and a single possible good consequence of it:

Now, to speak of public envy. There is yet some good in public envy, whereas in private there is none. For public envy is as an ostracism, that eclipseth men when they grow too great. And therefore it is a bridle also to great ones, to keep them within bounds. [VI, 396]

The essay closes with a warning of the force and prejudice of envy, and its power to taint all business. This is put extremely strongly, in contrast to the balanced temper of the civil essays:

It is also the vilest affection, and the most depraved; for which cause it is the proper attribute of the devil, who is called *The envious man, that soweth tares amongst the wheat by night*; as it always cometh to pass, that envy worketh subtilly, and in the dark; and to the prejudice of good things, such as is the wheat. [VI, 397]

Setting aside the digression (as it amounts to) on public envy, the illustrative and supporting devices in this essay all support a strong moral theme. The biased description (weighted by well-chosen examples) of circumstances under which envy is stimulated provides information about the ways in which the affection is aroused and acts, and at the same time persuades of its undeniable viciousness.[1] The essay is overtly condemnatory of

[1] It is difficult to see how Zeitlin judges this essay to be one in which 'the operations of envy are analyzed with a great deal of psychological acuteness but with no intimation of approval or disapproval'. See J. Zeitlin, 'The development of Bacon's Essays – with special reference to the question of Montaigne's influence upon them', *Journal of English and Germanic Philology*, 27 (1928), pp. 496–512; p. 512.

envy; at the same time it analyses extremely shrewdly the conditions which provoke envy, and hence gives convincing grounds for predicting situations likely to rouse envy in particular individuals.

This sort of presentational strategy is consistent with Bacon's remarks in the *De Augmentis* on the need for *Serious Satire* or the *Treatise of the Inner Nature of Things* [V, 18], for promoting virtuous behaviour in social situations. Bacon believed that if individuals are to behave 'dutifully' towards one another in a community they must be able to anticipate the possibilities for vice around them. The moral teacher should therefore teach his students about the 'frauds, cautions, impostures, and vices of every profession; for corruptions and vices are opposed to duties and virtues' [V, 17]:

> For it is not possible to join the wisdom of the serpent with the innocence of the dove, except men be perfectly acquainted with the nature of evil itself; for without this, virtue is open and unfenced; nay, a virtuous and honest man can do no good upon those that are wicked, to correct and reclaim them, without first exploring all the depths and recesses of their malice. [V, 17]

Once again, we are concerned here with *communication* of observations about the ways in which vices occur, and their effects, appropriately coloured to excite dislike and encourage avoidance. *Investigation* of the affections, as I have stressed, was part of the science of humanity, and Bacon's tentative investigations are to be found in the *Sylva Sylvarum*. For Bacon, moral teaching consists of instructions on how to behave dutifully or virtuously, presented as appealingly as possible. That is, communicated so as to prompt that assent and desire to act virtuously which cannot be produced by rational means:

> If the affections themselves were brought to order, and pliant and obedient to reason, it is true there would be no great use of persuasions and insinuations to give access to the mind, but naked and simple propositions and proofs would be enough. But the affections do on the contrary make such secessions and raise such mutinies and seditions...that reason would become captive and servile, if eloquence of persuasions did not win the imagination from the affections' part, and contract a confederacy between the reason and imagination against them. [IV, 456–7]

Moral teaching is not under normal circumstances concerned with teaching individuals to judge for themselves whether conduct is virtuous or vicious (since this is largely beyond them), but only to identify situations in which particular conduct is appropriate. The essay 'Of Envy' fits Bacon's pattern for preparing individuals for their encounters with vice, as part of the 'framing and predisposing of the minds of particular persons towards the preservation of [the] bonds of society' [V, 18]. That some at least of these later essays were designed as part of the *Treatise of the Inner Nature of Things* [V, 18] is suggested by the fact that the Latin translation of the 1625 *Essays* was entitled: *Sermones Fideles sive Interiora Rerum* [VI, 369]. The essay 'Of Envy' is a moral essay in the specific sense that by its method of presentation it plays a particular rôle in Bacon's plan for reforming ethical teaching.

I conclude by examining an essay which has customarily been regarded as an example of the virtuoso use of rhetoric for its own sake – an amusing exercise in equivocation. 'Of Truth' takes on a different aspect, I suggest, if viewed as an example of 'wisdom of transmission' in the field of ethics.

The essay opens with a strictly rhetorical 'entrance', an arresting and unspecific 'sentence' relating to the general topic 'truth':

> *What is Truth?* said jesting Pilate; and would not stay for an answer. [VI, 377]

This is followed, as if the transition were natural, by a specification of a particular sort of deviation from the truth: a refusal on the part of some men to settle on first principles in their speculations about true (that is, certain) knowledge. The specification is achieved by juxtaposing observations about such men, in the form of faintly disparaging comments associating 'mutability' in principles with 'giddiness' or unsoundness, and with lack of determination. This is summed up as an unwillingness on the part of such men to submit to the constraint which truth imposes on arbitrary speculation, and to face up to the rigour of the search for truth.

The summarising sentence carries the reader into the next stage in the essay. Men basically *enjoy* lies:

> But it is not only the difficulty and labour which men take in finding out of truth; nor again that when it is found it imposeth upon men's

thoughts; that doth bring lies in favour; but a natural though corrupt love of the lie itself. [VI, 377]

The early stages of the sentence attribute scientific unsoundness in principles to weakness of will and intellect, by innuendo. The conclusion of the sentence is that apart from being deterred by the demanding nature of the search for truth, men's flawed natures are actually pleased with lies. The transition is not reasoned, but observed, as if there were an evident link, in addition to the grammatical linking of clauses ('But not only ...nor again...but...').[1]

From this opening Bacon launches into a series of *sententiae* and examples which support by accumulation the assertion that men find lies more attractive than the truth. The initial specification of scientific truth is surreptitiously converted into that of truth as occasional lying – day-to-day misrepresentation of facts:

Doth any man doubt, that if there were taken out of men's minds vain opinions, flattering hopes, false valuations, imaginations as one would, and the like, but it would leave the minds of a number of men poor shrunken things, full of melancholy and indisposition, and unpleasing to themselves? [VI, 377]

The observation that unrelenting truthfulness in appraisal of a man's situation would produce lack of confidence, and neurotic self-disparagement, derives its force from the bias of the preceding illustrations. These represent truth as a pure and natural object, in contrast to lying, which is artificially brilliant and exciting:

But I cannot tell: this same truth is a naked and open day-light, that doth not shew the masks and mummeries and triumphs of the world, half so stately and daintily as candle-lights. Truth may perhaps come to the price of a pearl, that sheweth best by day; but it will not rise to the price of a diamond or carbuncle, that sheweth best in varied lights. A mixture of a lie doth ever add pleasure. [VI, 377]

The next transition, from the occasional untruth to 'living a lie', is once again made by a bridging sentence which apparently (and illegitimately) links the two specifications of 'untruth':

[1] It is probably the common tactic of using accumulation of examples, rather than the syllogism, to make points that has led critics to remark on similarity between Bacon's *Essays* and those of Montaigne. The similarity does not seem to me to go any deeper than this. On Bacon and Montaigne see Zeitlin, art. cit.

But it is not the lie that passeth through the mind, but the lie that sinketh in and settleth in it, that doth the hurt; such as we spake of before. [VI, 378]

'Such as we spake of before' does not in fact refer back to 'lying in first principles' in general, but to lying in ethical principles, that is, moral duplicity. The tone of this section is more ponderous, and Bacon uses biblical and classical citations to support the argument that truth is ethically equivalent to 'the good':

But howsoever these things are thus in men's depraved judgments and affections, yet truth, which only doth judge itself, teacheth that the inquiry of truth, which is the love-making or wooing of it, the knowledge of truth, which is the presence of it, and the belief of truth, which is the enjoying of it, is the sovereign good of human nature. [VI, 378]

This amounts to an expansion of the sentence given under the antitheses on 'Knowledge' in the *De Augmentis*:

All depraved affections are but false estimations; and gooodness and truth are the same thing. [IV, 482]

And it incorporates a version of Bacon's view that the 'sovereign good' of all natural things is 'in enjoying or fruition, effecting or operation, consenting or proportion, and approach or assumption' [III, 230]. In the extended version the tautologous manipulations of the term 'truth' add apparent depth and seriousness to the observation. The section culminates in another weighty and 'incontrovertible' sentence:

Certainly, it is heaven upon earth, to have a man's mind move in charity, rest in providence, and turn upon the poles of truth. [VI, 378]

The development so far discussed is contained within a single extended paragraph. In this way the careful transition and shifts of emphasis are made unobtrusive. The final section of the essay is contained in a separate paragraph, and shifts overtly from the philosophical aspects of truth to truth in civil affairs – honest dealing. In this case *sententiae* support the position that although dishonesty may be effective in business it is to be condemned:

It will be acknowledged even by those that practice it not, that clear and round dealing is the honour of a man's nature; and that mixture of falsehood is like allay in coin of gold and silver, which may make the metal work the better, but embaseth it. [VI, 378]

The passage conveys a moral judgment, that dishonesty is to be condemned, whereas in the essays on civil instruction those in authority were explicitly instructed how to exploit dissimulation to political advantage. This is in keeping with the distinction on which Bacon insists in his treatment of ethics, between moral and civil instruction. Moral instruction on the duty of individuals in a community does not teach policy, but persuades individuals to behave so as to uphold the 'bonds of society' [V, 18].

The close of the essay is a formal 'conclusion'. It is a forceful statement of the viciousness of dishonesty, based on a pun on the word 'faith':[1]

Surely the wickedness of falsehood and breach of faith cannot possibly be so highly expressed, as in that it shall be the last peal to call the judgments of God upon the generations of men; it being foretold, that when Christ cometh, *he shall not find faith upon the earth.*

[VI, 379]

The full weight of righteous condemnation is brought to bear on 'dishonesty' by equating 'breaking faith' (breaking one's word in society) with falsehood or untruthfulness in all areas of life, and finally with 'breaking God's faith' (rejecting christianity).

The essay 'Of Truth' is a carefully constructed, persuasive account of various degrees of falsehood (such as one is likely to meet with in daily experience in society), presented in such a way that the author's condemnation of falsehood is clear and compelling. By using this method of presentation Bacon succeeds in both describing in detail circumstances and characteristics of untruthfulness, and in evaluating them, without obscuring the facts by the judgment. He does not pretend that falsehood brings about its own downfall, as a deterrent to untruthfulness; he shows how things *are*, but colours his account to convey his judgment in his description. I suggest that it is in this sense,

[1] The same pun underlies the apophthegm recorded by Bacon on Henry IV of France, in his collected *Apophthegms*: 'King Henry the fourth of France was so punctual of his word, after it was once passed, that they called him *The King of the Faith*' [VII, 167]. Henry IV fought several bloody wars, and employed strategies of deceit and cunning, for the privilege of the *religious* title.

and not as an ethical pragmatist, that Bacon commends Machiavelli, and all those who 'openly and unfeignedly declare or describe what men do, and not what they ought to do' [V, 17].

It appears to be of the very nature of the *Essays*, and the determining factor in dictating the 'method' of their presentation, to employ what Bacon calls 'imaginative' or 'insinuative' reason [III, 383]. The individual essays are built up out of the devices which Bacon believed to make non-rational appeal, and to sway the reader's imagination, into a 'method of discourse' which ensures a favourable reception for the knowledge which they communicate. It is not necessary, indeed, perhaps not advisable, to show the reader the rational stages by which judgments in ethics and civics are arrived at. What is essential is to convince the reader to adopt spontaneously advised courses of action in particular circumstances. The complete absence of reasoned justification does not make Bacon an ethical or political pragmatist. Appeal to reason is not part of the strategy of the *Essays*.

APPENDIX

The terminology of ascending and descending methods

The purpose of this appendix is to collect together for convenient reference the various methods mentioned in the first chapter.

The idea that methods come in pairs – one an ascending method, the other a descending method (with sometimes a third, which neither ascends nor descends, added) – is fundamental to almost all renaissance writing on method. One of a pair descends from what is prior or better known to (or in) nature, to what is prior or better known to us. The other ascends from what is prior or better known to us, to what is prior or better known in nature. What is more general and abstract is prior in nature to what is individual and particular. The individual or particular is better known to us (because closer to our senses). At the cost of deliberate over-simplification one can explain the motive for this distinction as follows. The natural order of priority is the order of logical priority. Given the more general statement we can deduce a less general consequence, but not vice versa. However, it is particular instances which are evident to our senses, and in the process of acquisition of knowledge perception of particular instances precedes full understanding of their causes, and of universals. The order of priority relative to us is therefore not a logical order but an epistemological order. Ramus is exceptional in the period in refusing to accept any but natural priority (logical priority) as of any consequence to philosophy. The table on page 250 gives the terminology for the various discussions of ascending and descending method.

Unfortunately, the terminology of ascending and descending methods is not as straightforward as at first sight it appears. Indeed, much of the controversy about method in the dialectic handbooks and elsewhere is strictly on etymological issues. In the context of the present study the most important source of confusion lies in the fact that the term 'compositio' occurs both as an ascending method in the contexts of whole/parts and genera/species, and as a descending method in the context cause/effect. The primitive meanings of 'compositio' and 'resolutio' (as of 'synthesis' and 'analysis') are apparently those of the

	whole/parts	genera/species definition	cause/effect demonstration	axioms/theorems axiomatic
ASCENT	compositio synthesis	synthesis collectio compositio	τὸ ὅτι resolutive syllogism demonstratio quia demonstratio quod est analysis	a posteriori analysis resolutio
DESCENT	dissolutio resolutio partitio divisio analysis dissectio	analysis divisio diaeresis	τὸ διότι demonstratio potissima compositive syllogism demonstratio propter quid synthesis	a priori synthesis compositio
		definitio diaeresis	demonstratio potissima demonstratio simpliciter	reductio ad absurdum

whole/parts usage, and are carried over into the context of definition (genera/species) on the strength of the much debated analogy between the relation of whole to parts and the relation of genus to species. The contrary usage in demonstration (cause/effect) and axiomatic method (axioms/theorems) also derives from the primitive meanings of the terms. Just as in moving from whole to part we move from what is complex to what is simple (by resolution or analysis), so in moving from effect to cause or from theorem to axiom we move from what is complex and difficult to understand to what is simple. And just as in moving from simple parts to complex whole (by composition or synthesis) we move from what is simple to what is complex, so in moving from cause to effect or axiom to theorem we move from what is simple to what is complex. As I have already pointed out, it was a basic tenet of Aristotelianism that observed effects are intellectually more complex than their causes.

It is sometimes difficult in the renaissance to be sure when an author is genuinely confused in his use of traditional terminology, and when he is deliberately adapting it to a new purpose. Ramus, for example, deliberately uses the terminology of Platonic division for what is in fact an artificial method of partition. Digby, on the other hand, is accurately mocked by Temple for terminological confusion on the subject of method.

BIBLIOGRAPHY

WORKS BEFORE 1750

Agricola, R. *De Inventione Dialectica libri tres* ([Colonie], 1528)
Aphthonius *Progymnasmata Aphthonii*...(Lipsiae, 1591)
Averroës, *Aristotelis de physico auditu libri octo cum Averrois Cordubensis variis in eosdem commentariis*... (Venetiis, 1562)
Bacon, R. *De Mirabil. Potest. Artis et Naturae* (Parisiis, 1542)
Billingsley, H. *The Elements of Geometrie of the most auncient Philosopher Euclide of Megara* (London, 1570)
Blundeville, T. *The True Order and Method of Wryting and Reading Histories* (London, 1574)
Boccaccio, G. *De genealogiis deorum gentilium* (Basilea, 1532)
Bodin, J. *Methodus ad Facilem Historiarum Cognitionem* (Parisiis, 1566)
Caesarius, J. *Dialectica Joannis Caesarii...nunc recens Hermanni Raiiani Welsdalii fructuosis scholiis...illustrata*... (Coloniae Agrippinae, 1568)
Rhetorica (Coloniae, 1565)
Campanella, T. *Universalis Philosophiae* (Parisiis, 1638)
Cartari, V. *Le Imagini con la spositione dei dei de gli antichi* (Venetia, 1556 and Venetia, 1580)
Case, J. *Summa veterum interpretum in universam dialecticam Aristotelis* (Oxoniae, 1584)
Comes, N. *Mythologiae sive Explicationis Fabularum libri decem* (Venetiis, 1567 and Parisiis, 1583)
della Porta, G. B. *Magia naturalis* (Neapolis, 1589)
Digby, E. *De duplici methodo libri duo, unicam P. Rami methodum refutantes* (Londini, 1580)
Everardi Digbei Cantabrigiensis admonitioni Francisci Mildapetti responsio (Londini, 1580)
Theoria Analytica, viam ad Monarchiam Scientiarum demonstrans (Londini, 1579)
Eckius, J. *Joannis Eckii Theologi in Summulas Petri Hispani*...(Augustae Vindelicorum, 1516)
Elementarius Dialectice (Augustae Vindelicorum, 1517)

Erasmus, D. *Adagia* (Venetiis, 1508)
De duplici copia verborum ac rerum commentarii duo (Parisiis, 1512)
Parabolae sive similia (Argentorati, 1514)
Gentillet, I. *Discours sur le moyens de bien gouverner...un royaume...contre Machiavel* ([Genève], 1576)
Gocklenius, R. *Lexicon Philosophicum, quo tanquam clave philosophiae fores aperiuntur* (Francofurti, 1613)
Goveanus, A. *Antonii Goveani pro Aristotele responsio adversus Petri Rami calumnias* (Parisiis, 1543)
Hemmingius, N. *De lege naturae apodictica methodus* (Witebergae, 1577)
De methodis libri duo (Lipsiae, 1565)
Hispanus, P. *Expositio magistri Petri Tatareti in summulas Petri Hyspani* (Limoges, *c.* 1510)
Hooke, R. *Posthumous Works* ed. Waller (London, 1705)
Hoole, C. *A new discovery of the old arte of teaching schoole* (London, 1660)
Hoskins, J. *Directions for Speech and Style* (1599?) ed. Hudson (Princeton, 1935)
Louvain Academy *Commentaria, in Isagogen Porphyrii et in omnes libros Aristotelis de dialectica, olim...consilio et...sumptibus facultatis artium in...Academia Lovaniensi, per...peritissimos viros composita* (Lovanii, 1568)
Melanchthon, P. *De Dialectica* (Witebergae, 1531)
De rhetorica libri tres (Coloniae, 1521)
Dialectices libri quattuor in *Opera* (Basileae, 1541)
Erotemata Dialectices (Wittebergae, 1555)
Erotemata Dialectices (Witebergae, 1562)
παροιμίαι, *sive Proverbia Solomonis...cum adnotationibus Ph. Melanchthonis* (Haganoae, 1525)
Rhetoricae libri duo in *Opera* (Basileae, 1541)
Peacham, H. *The garden of eloquence* (London, 1593)
Powel, G. *Analysis Analyticorum Posteriorum sive librorum Aristotelis de Demonstratione* (Oxoniae, 1594)
Puttenham, G. *The Arte of English Poesie* (London, 1589)
Ramus, P. *Animadversionum Aristotelicarum libri XX* (Lutetiae, 1548)
Dialecticae Institutiones (Parisiis, 1543 and Lutetiae, 1547)
Scholae (Basileae, 1569)
Reisch, G. *Margarita Philosophica* (Friborgi [i.B], 1503)
Sanchez, F. *De multum nobili et prima universali scientia Quod Nihil Scitur* (Francofurti, 1618)
Sarcerius, E. *Dialectica* (Lipsiae, 1539)
Savonarola, G. *Compendium totius philosophiae, tam naturalis quam moralis* (Venetiis, 1534)

Scheckius, J. *P. Rami...et A. Talaei collectaneae prefationes, epistolae, orationes cum indice totius operis* (Parisiis, 1577 and Marpurgi, 1599)
De Demonstratione libri XV (Basileae, 1564)
Seton, J. *Dialectica Joannis Setoni Cantabrigiensis* (Londini, 1584)
Sprat, T. *History of the Royal Society* (London, 1667)
Swift, J. *A Tale of a Tub* ed. Davis (Oxford, 1939)
Temple, W. *Admonitio de unica P. Rami methodo reiectis caeteris retinenda* (Londini, 1580)
P. Rami Dialecticae libri duo, scholiis G. Tempelli Cantabrigiensis illustrati (Cantabrigiae, 1584)
Pro Mildapetti de unica methodo defensione contra Diplodophilum (Londini, 1581)
Trapezuntius, G. *Rhetoricorum libri V* (Venetiis, 1523)
Valla, L. *Laurentii Vallae Romani, Dialecticarum disputationum libri tres...*(Coloniae, 1541)
Watts, G. *Of the advancement and proficience of learning or the partition of sciences IX bookes...Interpreted by G. Wats* (Oxford, 1640)
Willichius, J. *De methodo omnium artium et disciplinarum...*(Francof. ad Viadrum, 1550)
Wilson, T. *The Arte of Rhetorique* (London, 1560)
The Rule of Reason (London, 1551)
Zabarella, J. *Jacobi Zabarellae Patavini opera logica* (Francofurti, 1608)

WORKS AFTER 1750

Anderson, F. *The Philosophy of Francis Bacon* (Chicago, 1948)
Anglo, S. *Machiavelli: A Dissection* (London, 1969)
Arber, A. *Herbals: Their Origin and Evolution* (Cambridge, 1953)
Arber, E. *A Harmony of the Essays...of Francis Bacon* (London, 1871)
Bailey, C. *Lucretius' De Rerum Natura* (Oxford, 1947)
Baldwin, T. W. *William Shakspere's small Latine and lesse Greeke* (Urbana, 1944)
Baur, L. 'Dominicus Gundissalinus *De Divisione Philosophiae:* herausgegeben und philosophiegeschichtlich untersucht nebst einer Geschichte der philosophischen Einleitung bis zum Ende der Scholastik', *Beiträge zur Geschichte der Philosophie des Mittelalters* ed. Baeumker and von Hertling, Band IV, Heft 2–3 (Münster, 1903)
Bird, O. 'The tradition of logical topics: Aristotle to Ockham', *Journal of the History of Ideas,* 23 (1962), pp. 307–23
Blake, R. M. 'Theory of Hypothesis among Renaissance Astronomers', in Madden, Blake & Ducasse, *Theories of Scientific Method* (Washington, 1960), pp. 22–49
Blanchet, L. *Campanella* (Paris, 1920)

Blunt, H. W. 'Bacon's method of science', *Proceedings of the Aristotelian Society*, 4 (1903–4), pp. 16–31

Bolgar, R. R. *The Classical Heritage and its Beneficiaries* (Cambridge, 1958)

Bowman, M. A. *The English Prose Style of Sir Francis Bacon* (unpublished Ph.D. diss., Wisconsin, 1964)

Brett, G. S. *A History of Psychology* (London, 1921) vol. II

Broad, C. D. *The Philosophy of Francis Bacon* (Cambridge, 1926) reprinted in *Ethics and the History of Philosophy* (London, 1952)

Burke, P. 'A survey of the popularity of ancient historians, 1450–1700', *History and Theory*, 5 (1966), pp. 135–52

Burke, P. B. 'Rhetorical considerations of Bacon's style', *College Composition and Communication*, 18 (1967), pp. 23–31

Busch, W. *England under the Tudors, vol. I: King Henry VII*, transl. Todd (London, 1895)

Carman, B. E. *A Study of Natalis Comes' theory of mythology and its influence in England* (unpublished Ph.D. diss., London, 1970)

Carozzi, A. V. 'New historical data on the origin of the theory of continental drift', *Geological Society of America Bulletin*, 81 (1970), pp. 283–6

'A propos de l'origine de la théorie des dérives continentales: Francis Bacon (1620), François Placet (1668), A. von Humboldt (1801) et A. Snider (1858)', *Compte rendu des séances de la société de physique et d'histoire naturelle de Genève*, 4 (1969), pp. 171–9

Chabod, F. *Machiavelli and the Renaissance* (Harvard, 1958)

Clark, D. L. *Rhetoric in Greco-Roman education* (New York, 1957)

Clark, D. S. T. *Francis Bacon: The Study of History and the Science of Man* (unpublished Ph.D. diss., Cambridge, 1970)

Cochrane, R. C. 'Francis Bacon and the use of the mechanical arts in eighteenth century England', *Annals of Science*, 12 (1956), pp. 137–56

Costello, W. T. *The Scholastic Curriculum at Early Seventeenth Century Cambridge* (Harvard, 1958)

Crane, R. S. 'The relation of Bacon's *Essays* to his programme for the advancement of learning', *Schelling Anniversary Papers* (New York, 1923), pp. 87–105, reprinted in Vickers ed., *Essential Articles for the study of Francis Bacon* (Connecticut, 1968), pp. 272–92

Crane, W. G. (ed.) *Peacham's Garden of Eloquence (1593)* (Florida, 1954)

Crescini, A. *Le origini del metodo analitico: il cinquecento* (Udine, 1965)

Il problema metodologico alle origini della scienza moderna (Rome, 1972)

Crombie, A. C. *Robert Grosseteste and the origins of experimental science 1100–1700* (Oxford, 1953)

Curtis, M. H. *Oxford and Cambridge in Transition: 1558–1642* (Oxford, 1959)

Dassonville, M-M. *Pierre de la Ramée, Dialectique (1555)* (Geneva, 1964)

Davis, W. R. 'The imagery of Bacon's late work', *Modern Language Quarterly*, 27 (1966), pp. 162–73

Dean, L. F. 'Francis Bacon's theory of civil history-writing', *English Literary History*, 8 (1941), pp. 161–83, reprinted in Vickers ed., *Essential Articles for the study of Francis Bacon* (Connecticut, 1968), pp. 211–35

de Mas, E. 'L'origine della norma e della sanzione giuridica nel pensiero di Francesco Bacone', *Rivista Internazionale di Filosofia del Diritto*, 37 (1960), pp. 143–9

de Rijk, L. M. (ed.) *Tractatus, called afterwards Summule Logicales...* (Assen, 1972)

Donahue, W. H. *The Dissolution of the Celestial Spheres* (unpublished Ph.D. diss., Cambridge, 1972)

Ducasse, C. J. 'Francis Bacon's philosophy of science', *Structure, Method and Meaning* ed. Henle *et al.* (New York, 1951), pp. 115–44

Dunbar, H. F. *Symbolism in Mediaeval Thought* (Yale, 1929)

Eucken, R. *Geschichte der philosophischen Terminologie* (Hildesheim, 1964)

Farrington, B. *Francis Bacon: Philosopher of Industrial Science* (New York, 1949)

Finley, M. I. 'Myth, memory and history', *History and Theory*, 4 (1964–5), pp. 281–302

Flint, R. *Philosophy as Scientia Scientiarum, and A History of Classifications of the Sciences* (Edinburgh and London, 1904)

Foster Watson *The English Grammar Schools to 1660: their Curriculum and Practice* (Cambridge, 1908)

Freudenthal, J. 'Beiträge zur Geschichte der englischen Philosophie [I]', *Archiv für Geschichte der Philosophie*, 4 (1891), pp. 450–77

'Beiträge zur Geschichte der englischen Philosophie [III]', *Archiv für Geschichte der Philosophie*, 5 (1892), pp. 1–41

Fueter, E. *Geschichte der neuern Historiographie* (Munich/Berlin, 1936)

Fussner, F. S. *The Historical Revolution: English Historical Writing and Thought 1580–1640* (London, 1962)

Garin, E. *L'umanesimo italiano* (Bari, 1965)

Garner (Carman), B. E. 'Francis Bacon, Natalis Comes and the mythological tradition', *Journal of the Warburg and Courtauld Institutes*, 33 (1970), pp. 264–97

Gilbert, F. *Machiavelli and Guicciardini: Politics and History in sixteenth-century Florence* (Princeton, 1965)

Gilbert, N. W. *Renaissance Concepts of Method* (New York, 1960)

Gilmore, M. P. *Humanists and Jurists: six studies in the Renaissance* (Harvard, 1963)

Gooder, R. D. *The Rhetorical Work of Juan Luis Vives* (unpublished Ph.D. diss., Cambridge, 1969)

Hall, A. R. *The Scientific Revolution 1500–1800* (London, 1954)

Hallward, B. L. 'Cicero historicus', *Cambridge Historical Journal*, 3 (1931), pp. 221–37

Harvey, E. R. *The Inward Wits: An Enquiry into the Aristotelian Tradition in Faculty Psychology in its literary relations during the later middle ages and the Renaissance* (unpublished Ph.D. diss., London, 1970)

Heath, T. 'Logical grammar, grammatical logic, and humanism in three German Universities', *Studies in the Renaissance*, 18 (1971) pp. 9–64

Hesse, M. B. 'Francis Bacon's philosophy of science', *A Critical History of Western Philosophy* ed. O'Connor (New York, 1964), reprinted in Vickers ed., *Essential Articles for the study of Francis Bacon* (Connecticut, 1968), pp. 114–39

Howell, W. S. *Logic and Rhetoric in England 1500–1700* (Princeton, 1956)

The Rhetoric of Alcuin and Charlemagne (Princeton, 1941)

Jaeger, H. 'Introduction aux rapports de la pensée juridique et de l'histoire des idées en Angleterre, depuis la Réforme jusqu'au XVIII siècle', *Les Archives de Philosophie du droit*, 15 (1970), pp. 13–70

Jardine, L. A. 'The place of dialectic teaching in sixteenth century Cambridge', *Studies in the Renaissance* (1974)

Kearney, H. *Scholars and Gentlemen: Universities and Society in Pre-Industrial Britain, 1500–1700* (London, 1970)

Kennedy, G. A. *The Art of Persuasion in Greece* (London, 1963)

King, D. B. & Rix, H. D. (ed.) *On copia of words and ideas* (Wisconsin, 1963)

Kneale, W. and M. *The Development of Logic* (Oxford, 1962)

Kocher, P. H. 'Francis Bacon on the science of jurisprudence', *Journal of the History of Ideas*, 18 (1957), pp. 3–26, reprinted in Vickers ed., *Essential Articles for the study of Francis Bacon* (Connecticut, 1968), pp. 167–94

Kosman, L. A. *The Aristotelian Backgrounds of Bacon's 'Novum Organum'* (unpublished Ph.D. diss., Harvard, 1964)

Kristeller, P. O. *Renaissance Thought* (New York, 1961)

Renaissance Thought II (New York, 1965)

Renaissance Concepts of Man (New York, 1972)

Lasswitz, K. *Geschichte der Atomistik vom Mittelalter bis Newton* (Hildesheim, 1963), first edition (Hamburg and Leipzig, 1890)

Lee, H. D. P. 'Geometrical method and Aristotle's account of first principles', *Classical Quarterly*, 29 (1935), pp. 113–24

Lemmi, C. W. *The Classical Deities in Bacon: A Study in Mythological Symbolism* (Baltimore, 1933)

Levi, A. *Il pensiero di F. Bacone considerato in relazione con le filosofie della natura del Rinascimento e col razionalismo cartesiano* (Turin, 1925)
Levy, F. J. *Tudor Historical Thought* (San Marino, 1967)
Lewis, C. S. *The Literary Impact of the Authorised Version* (London, 1950)
Lloyd, G. E. R. *Aristotle: the growth and structure of his Thought* (Cambridge, 1965)
Luciani, V. 'Bacon and Guicciardini', *Proceedings of the Modern Languages Association,* 62 (1947), pp. 96–113
'Bacon and Machiavelli', *Italica,* 24 (1947), pp. 26–40
Francesco Guicciardini and his European Reputation (New York, 1936)
McCabe, B. 'Francis Bacon and the natural law tradition', *Natural Law Forum,* (1964), pp. 111–21
Macciò, M. 'A proposito dell'atomismo nel "Novum Organum" di Bacone', *Rivista critica di storia della filosofia,* 17 (1962), pp. 188–96
McKeon, R. 'Aristotle's conception of the development and the nature of scientific method', *Journal of the History of Ideas,* 8 (1947), pp. 3–44
Maitland, S. R. 'Archbishop Whitgift's accounts', *British Magazine,* 32, 33 (1847, 1848), *passim*
Mandeville, S. *The Rhetorical Tradition of the Sententia; with a Study of its Influence on the Prose of Sir Francis Bacon and of Sir Thomas Browne* (unpublished Ph.D. diss., St. Louis, 1960) (abstract)
Morrow, G. R. *Proclus, A commentary on the First Book of Euclid's Elements* (Princeton, 1970)
Mosher, J. H. *The exemplum in early religious and didactic literature in England* (New York, 1911)
Mullally, J. P. *The Summulae Logicales of Peter of Spain* (Indiana, 1945)
Mullinger, J. B. *The University of Cambridge from the Royal Injunctions of 1535 to the accession of Charles the First* (Cambridge, 1884)
Nadel, G. H. 'History as psychology in Francis Bacon's theory of history', *History and Theory,* 5 (1966), pp. 275–87, reprinted in Vickers ed., *Essential Articles for the study of Francis Bacon* (Connecticut, 1968), pp. 236–50
Olschki, L. *Machiavelli the Scientist* (California, 1945)
Ong, W. J. *Ramus and Talon Inventory* (Harvard, 1958)
Ramus, Method, and the Decay of Dialogue (Harvard, 1958)
Orsini, G. N. G. *Bacone e Machiavelli* (Genoa, 1936)
Owen, G. E. L. 'Tithenai ta Phainomena', *Aristotle: A Collection of critical Essays* ed. Moravcsik (London, 1968), pp. 167–90
Paetow, L. J. *The Arts Course at Medieval Universities with special reference to Grammar and Rhetoric* (Illinois, 1910)
Paré, G. M., Brunet, A. & Tremblay, P. *La renaissance du XIIe siècle* (Paris, 1933)
Pegis, A. C. *Basic Writings of Saint Thomas Aquinas* (New York, 1944)

Pineas, R. 'The polemical *exemplum* in sixteenth century religious controversy', *Bibliothèque d'Humanisme et Renaissance,* 28 (1966), pp. 393–6

'John Frith's polemical use of rhetoric and logic', *Studies in English Literature,* 4 (1964), pp. 85–100

Popkin, R. H. *The History of Scepticism from Erasmus to Descartes* (Assen, 1960)

Praz, M. *Machiavelli and the Elizabethans* (London, 1928)

Prest, W. R. *The Inns of Court under Elizabeth I and the Early Stuarts 1590–1640* (London, 1972)

Primack, M. *Francis Bacon's Philosophy of Nature, and Teleology and Mechanism in the Philosophy of Francis Bacon* (unpublished Ph.D. diss., Johns Hopkins, 1962)

'Outline of a reinterpretation of Francis Bacon's philosophy', *Journal of the History of Philosophy,* 5 (1967), pp. 123–32

Prior, M. E. 'Bacon's man of science', *Journal of the History of Ideas,* 15 (1954), pp. 348–70, reprinted in Vickers ed., *Essential Articles for the study of Francis Bacon* (Connecticut, 1968), pp. 140–63

Raab, F. *The English Face of Machiavelli* (London, 1964)

Randall, J. H. 'The development of scientific method in the school of Padua', *Journal of the History of Ideas,* 1 (1940), pp. 177–206, reprinted in *The School of Padua and the emergence of Modern Science* (Padua, 1961), pp. 15–68

Rashdall, H. *Rashdall's Mediaeval Universities* ed. Powicke and Emden (Oxford, 1936), vol. III

Reif, P. 'The textbook tradition in natural philosophy, 1600–1650', *Journal of the History of Ideas,* 30 (1969), pp. 17–32

Renaudet, A. *Machiavel* revised edition (Paris, 1956)

Reynolds, B. *Method for the easy comprehension of history* (New York, 1945)

Risse, W. *Die Logik der Neuzeit* Band I (Stuttgart, 1964)

Rossi, P. *Clavis Universalis* (Milan, 1960)

Francesco Bacone, dalla magia alla scienza (Bari, 1958), transl. Rabinovitch (London, 1968)

I filosofi e le macchine (1400–1700) (Milan, 1962)

Schmitt, C. B. *A Critical Survey and Bibliography of Studies on Renaissance Aristotelianism 1958–1969* (Padua, 1971)

'Experience and experiment: a comparison of Zabarella's view with Galileo's in *De Motu*', *Studies in the Renaissance,* 16 (1969), pp. 80–138

'Experimental evidence for and against a void: the sixteenth century arguments', *Isis,* 58 (1967), pp. 352–66

Cicero Scepticus: a study of the influence of the Academica in the Renaissance (The Hague, 1972)

Schneiders, W. 'Einige Bemerkungen zum gegenwärtigen Stand der Bacon-Forschung', *Zeitschrift für Philosophische Forschung,* 16 (1962), pp. 450–71

Schüling, H. *Die Geschichte der axiomatischen Methode im 16. und beginnenden 17. Jahrhundert* (Hildesheim, 1969)

Sears, Jayne *Library Catalogues of the English Renaissance* (Los Angeles, 1956)

Simon, J. *Education and Society in Tudor England* (Cambridge, 1965)

Smalley, B. *The Study of the Bible in the Middle Ages* (Oxford, 1952)

Snow, R. E. *The Problem of certainty: Bacon, Descartes and Pascal* (unpublished Ph.D. diss., Indiana, 1967)

Sperling and Simon (ed.) *The Zohar* (London, 1949)

Stone de Montpensier, R. L. 'Bacon as lawyer and jurist', *Archiv für Rechts- und Sozialphilosophie,* 54 (1968), pp. 449–83

Strickland Gibson *Statuta Antiqua Universitatis Oxoniensis* (Oxford, 1931)

Tuve, R. *Elizabethan and Metaphysical Imagery* (Chicago, 1947)

van Deusen, N. C. *Telesio: The First of the Moderns* (New York, 1932)

van Leeuwen, H. *The problem of certainty in English thought, 1630–90* (The Hague, 1963)

Vasoli, C. *La dialettica e la retorica dell'Umanesimo* (Milan, 1968)

Viano, C. A. 'Esperienza e natura nella filosofia di Francesco Bacone', *Rivista di Filosofia* (1954), pp. 291–313

Vickers, B. W. 'Bacon's use of theatrical imagery', *Studies in the Literary Imagination,* 4 (1971), pp. 189–226

Classical Rhetoric in English Poetry (London, 1970)

Essential Articles for the study of Francis Bacon (Connecticut, 1968)

Francis Bacon and Renaissance Prose (Cambridge, 1968)

von Fritz, K. 'Die ἐπαγωγή bei Aristoteles', *Sitzungsberichte der Bayerischen Akademie der Wissenschaften* (1964)

von Leyden, W. *Seventeenth Century Metaphysics* (London, 1968)

von Sigwart, C. *Logik* (Tübingen, 1873), transl. Dendy (London, 1890)

Walker, D. P. 'Francis Bacon and *spiritus*', *Science, Medicine and Society in the Renaissance* ed. Debus (New York, 1972), vol. II, pp. 121–130.

Spiritual and Demonic Magic from Ficino to Campanella (London, 1958)

Wallace, K. *Francis Bacon on the Nature of Man* (Illinois, 1967)

Watson, G. *The Stoic Theory of Knowledge* (Belfast, 1966)

Weisheipl, J. A. 'Classification of the sciences in medieval thought', *Mediaeval Studies,* 27 (1965), pp. 54–90

'Curriculum of the faculty of arts at Oxford in the early fourteenth century', *Mediaeval Studies* 26 (1964), pp. 143–85

'The place of the liberal arts in the university curriculum during the XIVth and XVth centuries', *Actes du Quatrième Congrès International de Philosophie Mediévale* (Montreal, 1969)

Welter, J. T. *L'exemplum dans la littérature religieuse et didactique du moyen âge* (Paris, 1927)

Wheeler, T. V. H. *Sir Francis Bacon as Historian* (unpublished Ph.D. diss., N. Carolina, 1955)

'Bacon's purpose in writing *Henry VII*', *Studies in Philology*, 54 (1957), pp. 1–13

Whitaker, V. K. *Bacon and the Renaissance Encyclopedists* (unpublished Ph.D. diss., Stanford, 1933)

Wightman, W. P. D. 'Quid sit Methodus?', *Journal of the History of Medicine and Allied Sciences*, XIX (1964), pp. 360–76

'Les problèmes de methode dans l'enseignment medical à Padoue et à Ferrare', *Sciences de la Renaissance* (Paris, 1973), pp. 187–95

Yates, F. *The Art of Memory* (London, 1966)

Zeitlin, J. 'The development of Bacon's Essays – with special reference to the question of Montaigne's influence upon them', *Journal of English and Germanic Philology*, 27 (1928), pp. 496–512

INDEX